# Glycerine Production and Transformation - An Innovative Platform for Sustainable Biorefinery and Energy

*Edited by Marco Frediani, Mattia Bartoli and Luca Rosi*

Published in London, United Kingdom

IntechOpen

*Supporting open minds since 2005*

Glycerine Production and Transformation -
An Innovative Platform for Sustainable Biorefinery and Energy
http://dx.doi.org/10.5772/intechopen.78821
Edited by Marco Frediani, Mattia Bartoli and Luca Rosi

Contributors
Marco Almeida, Josimari Paschoaloto, José Luis Contreras, Israel Pala-Rosas, Jose Salmones, Beatriz Zeifert, Jose Luis Contreras, Priya Samudrala, Masahide Yasuda, Toshiaki Yamashita, Tomoko Matsumoto, Vinicius Rossa, Germildo Juvenal Muchave, Gisel Chenard Díaz, Sibele B. C. Pergher, Donato Alexandre Gomes Aranda, Jun Wang, Linlin Zhu, Shulin Chen, Jinzheng Wang, Yan Hu, Marco Frediani, Luca Rosi, Mattia Bartoli



First published in London, United Kingdom, 2019 by IntechOpen
IntechOpen is the global imprint of INTECHOPEN LIMITED, registered in England and Wales, registration number: 11086078, The Shard, 25th floor, 32 London Bridge Street
London, SE19SG - United Kingdom
Printed in Croatia

British Library Cataloguing-in-Publication Data
A catalogue record for this book is available from the British Library

Additional hard and PDF copies can be obtained from orders@intechopen.com

Glycerine Production and Transformation -
An Innovative Platform for Sustainable Biorefinery and Energy
Edited by Marco Frediani, Mattia Bartoli and Luca Rosi
p. cm.
Print ISBN 978-1-78984-690-4
Online ISBN 978-1-78984-691-1
eBook (PDF) ISBN 978-1-83962-179-6

For EU product safety concerns:
IN TECH d.o.o., Prolaz Marije Krucifikse Kozulić 3, 51000 Rijeka, Croatia, info@intechopen.com or visit our website at intechopen.com.

# Meet the editors

Marco Frediani is a researcher of industrial chemistry at the University of Florence, Department of Chemistry "Ugo Schiff." He obtained his PhD under the supervision of Dr. Claudio Bianchini (Institute of Chemistry and Organometallic Compounds at the National Research Council (CNR), Florence, Italy) and Prof. Dr. Walter Kaminsky (Institute of Technical and Macromolecular Chemistry, University of Hamburg). His main scientific interests are focused on the use of catalysts for organic reactions, polymer synthesis from oil and biomasses, and the reuse of end of life materials to obtain new chemical products and fuels by microwave pyrolysis (tires, biomasses, etc.).

Mattia Bartoli received a master's degree (MSc) in Chemical Science with honors at the University of Florence. In 2017 he received a PhD in Chemical Science from the University of Florence. After a period as a postdoc research fellow at the University of Alberta, he started work as a member of the carbon group with the qualification of research assistant at the Polytechnic University of Turin. He is a member of the Italian Chemical Society and INSTM consortium.

Luca Rosi is Associate Professor of Industrial Chemistry at the Department of Chemistry "Ugo Schiff" of the University of Florence. His scientific activity was initially focalized on studies in the field of the "homogeneous catalysis," with a particular interest in the reactivity of complexes of metals of VIII group elements (Ru, Co, Pd). Nowadays, his attention has been mainly focused on pyrolytic processes adopting microwaves for the treatment of end-of-life polymeric parts (e.g. tires, postconsumer plastics, plastic solid waste, shredded mix, WEEE) and/or biomass (waste) in useful products or fuels (value), as well as bio-oil and char from pyrolysis.

# Contents

# Preface

Biodiesel commodities have been the main driving forces of the sustainable economy. Their production is mainly obtained through transesterification of triglycerides using small chain alcohols, such as methanol or ethanol, leading to the massive production of crude glycerol as a by-product. Purification of crude glycerol to a chemically pure substance is costly and the glycerol market is already saturated. Consequently, the price of crude glycerol continues to decline and directly affect biodiesel production cost. In this scenario, glycerol has become an attractive molecule without a well-known industrial destiny but with an impressive yearly industrial surplus. Nonetheless, glycerol is one of the most polyhedral bioderived molecules for enforcing chemical and biochemical conversion platforms. The three hydroxylic functionalities on three different carbon atoms represent an astonishing scaffold for a wide range of chemical transformations in the intermediates and fine chemical production. Glycerol is the starting material for the synthesis of fuel additives (i.e. acetins) for multi-functionality acetals and compounds like 1,3-dihydroxypropanone or acroleic acid. Furthermore, neat glycerol is currently used for a number of applications ranging from the manufacturing of drugs (i.e. drug solvents, carriers for antibiotics and antiseptics, plasticizers for capsules) and medicine preparations (cough syrups, ear infection preparations). Moreover, glycerol has found several uses in the food and beverage industry as a solvent, sweetener, and preservative agent. Another principal glycerol use is related to energy production. In this field, great interest has been harvested in the production of hydrogen through a variety of different processes (i.e. steam reforming, partial oxidation, auto-thermal reforming). The current specialistic literature lacks a comprehensive approach to the world of glycerol use. Plenty of published works have reported a very narrow aspect of complex glycerol use and conversion reality. In this book, we decided to report a multidisciplinary approach that takes into account chemistry, engineering, and biological sciences. We are deeply persuaded that only an expanded point of view could clarify all the challenges and possibilities related to this exciting research field. Consequently, we promote the realization of a specialist book by different researchers. Through its chapters, this monographic work presents an overview of whether glycerol-based processes are convenient and economically sound considering both traditional and unconventional approaches. We are pleased to say that a number of specialists in different disciplines collaborated with us to provide the most exhaustive points of view on glycerol uses to all readers.

Ranging from more traditional to unconventional conversion routes, a wide range of studies has been collected to enlighten in a very clear way the realm of glycerine.

We hope that this book will be the future reference for all enthusiastic researchers who are contributing to the further development of glycerol valorization.

**Marco Frediani**
Department of Chemistry "Ugo Schiff,"
University of Florence,
Sesto Fiorentino, Italy

**Mattia Bartoli**
Department of Applied Science and Technology,
Polytechnic of Turin,
Turin, Italy

**Luca Rosi**
Department of Chemistry "Ugo Schiff,"
University of Florence,
Sesto Fiorentino, Italy

# Section 1

# General

Chapter 1

# Introductory Chapter: A Brief Insight about Glycerol

*Mattia Bartoli, Luca Rosi and Marco Frediani*

## 1. Toward sustainability: glycerol and integrated conversions

Over the last century, oil-based economy drove the mankind on the edge of environmental collapse [1, 2]. Moreover, the continuous increment of oil consumption has rapidly decreased the worldwide reserves [3]. During the last decades, the increased accountability of companies with regards to environmental issues has represented a formidable driving force for the development of sustainable industrial processes together with innovative products. As a consequence, the production of fuels and chemicals has started to use recycled or renewable feedstocks in place of oil-based raw materials attempting to improve process sustainability [4]. Biodiesel commodities have been of the main driving forces of the sustainable economy simultaneously decreasing the emission of carbon dioxide and producing affordable biofuels for worldwide market. Actually, biodiesel production is mainly performed through transesterification of triglycerides using small chains alcohols (i.e., methanol, ethanol, etc.) leading to the massive production of glycerol as byproduct. A general estimation based on the actual production evaluate for 2020 an annual surplus of glycerol around to $4.2 \times 10^6$ ton/year [5].

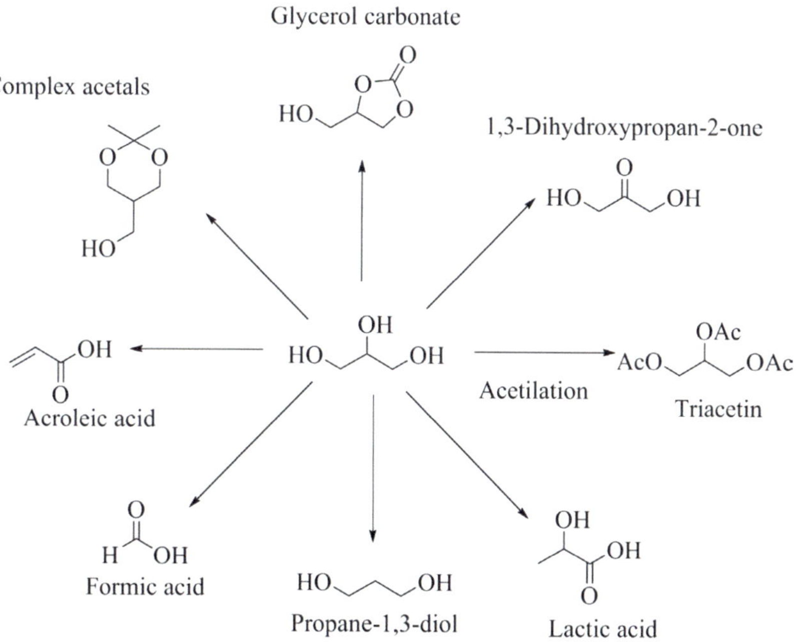


**Figure 1.**
*Glycerol as sound candidate for multiconversion platform for chemicals production.*

Despite its large availability, producing refined glycerol from those recovered in the biodiesel stream is far more expensive than common production cause the costs of purification process. Thus, glycerol becomes an attractive molecule without a well-known industrial destiny [6]. Many authors have recognized its value as feedstock for a lot of conversions (i.e., oxidation, hydrogenolysis, etherification, esterification, etc.) without proposing a breakthrough technology for its valorization. Despite this, glycerol processing is one of the most polyhedral bio-derived molecules for enforcing chemical and biochemical conversions platforms [7]. The great versatility of glycerol led to its incorporation in multi approach bio-based processes dedicated on obtaining several chemicals (i.e., glycerol carbonate, drugs synthones, etc.), **Figure 1**.

The astonishing possibilities of glycerol are still unexploited. Nonetheless, it could become the resource for further development in the field of both green chemistry and blue economy.

## 2. Approaching the readers

The simultaneous glycerol conversion requires multidisciplinary approach that takes into account chemistry, engineering, and biological sciences.

Available literature generally lacks in a comprehensive point of view, focusing more on one single aspect of the problem. In this book, we collected contributes that tries to paint a complete and multidisciplinary picture on the great possibilities related to glycerol.

Through the chapters, we reported the use of glycerine as solvent, food additive, monomer for textiles polymers, and drugs. We discuss a lot about successful achievements of converting glycerol into value-added products, using a lot of approaches to enlighten the technical and scientific issues and solutions of each of them.

Furthermore, we would like to give readers a handy and effective tool to easily understand how this field is interesting and diverse.

Through its chapters, this monographic opera presents an overview on whether glycerol-based processes are convenient and economically sound considering both traditional and unconventional approaches.

This book could be easily used by any reader with a strong scientific background ranging from scientific company advisors to academia members. Nonetheless, students enrolled in scientific undergraduate and graduate programs could be consulted to this text for any further and deeper investigation.

In the end, we proposed a very high scientific content book that could represent the reference text for any consideration and future study about glycerol for the next years.

## Author details

Mattia Bartoli[2], Luca Rosi[1] and Marco Frediani[1*]

1 Department of Chemistry "Ugo Schiff", University of Florence, Sesto Fiorentino, Italy

2 Department of Applied Science and Technology, Polytechnic of Turin, Turin, Italy

*Address all correspondence to: marco.frediani@unifi.it

## References

[1] Heinberg R. The Oil Depletion Protocol: A Plan to Avert Oil Wars, Terrorism and Economic Collapse. Canada: New Society Publishers; 2009. ISBN-13: 978-0865715639

[2] Newman P. Beyond peak oil: Will our cities collapse? Journal of Urban Technology. 2007;**14**:15-30

[3] Khisamov RS, Safarov AF, Kalimullin AM, Dryagalkina AA. Probabilistic-statistical estimation of reserves and resources according to the international classification SPE-PRMS. Part 1. Georesursy = Georesources. 2018;**20**(3):158-164. DOI: https://doi.org/10.18599/grs.2018.3.158-164

[4] Török B, Dransfield T. Green chemistry: Historical perspectives and basic concepts. In: Green Chemistry. The Netherland: Elsevier; 2018. pp. 3-16. ISBN: 9780128092705

[5] Monteiro MR, Kugelmeier CL, Pinheiro RS, Batalha MO, da Silva César A. Glycerol from biodiesel production: Technological paths for sustainability. Vol. 88(C). Renewable and Sustainable Energy Reviews. Elsevier; 2018. pp. 109-122

[6] Mota CJ, Pinto BP, De Lima AL. Glycerol: A Versatile Renewable Feedstock for the Chemical Industry. Springer: International Publishing; 2017

[7] Tan HW. Abdul Aziz AR, Aroua MK. Glycerol production and its applications as a raw material: A review. Vol. 27(C). Renewable and Sustainable Energy Reviews. Elsevier; 2013. pp. 118-127

Section 2

# Catalytic Conversion

Chapter 2

# Catalytic Dehydration of Glycerine to Acrolein

*Israel Pala Rosas, Jose Luis Contreras Larios, Beatriz Zeifert and José Salmones Blásquez*

## Abstract

The biodiesel production yields glycerine as a by-product in quantities around 10 vol% of produced biodiesel. Acrolein can be obtained from glycerine by a dehydration reaction. Catalytic processes in gas phase have been developed to obtain acrolein from a renewable feedstock using heterogeneous catalysts. The main process variables are the reaction temperature, the concentration of glycerol in water, and the space velocity in fixed-bed reactors. A thermodynamic study of the equilibrium has been made to estimate the conversion to equilibrium as a function of temperature. The reactors have been heated usually between 523 and 603 K. Generally, an aqueous glycerol solution is preheated in a preheating zone at a temperature enough to vaporize the feedstock, between 473 and 533 K, depending on the concentration of reactant required in the feed. Some of the most active catalysts in the gas-phase reaction (yield >70%) were $NH_4$-La-β zeolite, Pd/LaY zeolite, hierarchical ZSM-5, $WO_3/ZrO_2$, $WO_3/TiO_2$, ZrOx-NbOx, WOx-NbOx, $WO_3$-$SiO_2/ZrO_2$, NbOx-WOx/$Al_2O_3$, $H_3PO_4$-MCM-41, SAPO-40, NbPSi, Pd-$H_3PW_{12}O_{40}$/Zr-MCM-41, $H_3PW_{12}O_{40}$/Cs-SBA-15, $H_3PW_{12}O_{40}/Nb_2O_5$, Cs-doped $H_4SiW_{12}O_{40}$/$Al_2O_3$, $H_4SiW_{12}O_{40}/TiO_2$, and $H_4SiW_{12}O_{40}/SiO_2$.

**Keywords:** glycerine dehydration, acrolein, renewable production, acid catalyst

## 1. Introduction

During the last decades, the growing demand of energy and the depletion of fossil resources have resulted in the research and development of sustainable technologies for the production of valuable chemical compounds and fuels, and biomass conversion through catalytic processes is a potential alternative. One of the most viable choices for the partial replacement of petroleum diesel is the use of biodiesel as fuel in internal combustion engines. The biodiesel production yields glycerine (glycerol or 1,2,3-propanetriol) as by-product in quantities around 10% of the volume of produced biodiesel, and, as a result of the development of biodiesel industry, the global production of glycerine has increased while its market price has consequently declined [1].

From this perspective, intensive research has been carried out in recent years to develop biotechnological and catalytic processes that allow the change of the current status of glycerine as a by-product into a raw material for the production of compounds of industrial and technological interests [1, 2]. The catalytic dehydration of glycerine has become important because it may yield acrolein (2-propenal) as the main reaction product and represents a route for its renewable production, in contrast with the current process based on the partial oxidation of propylene derived from the petrochemical industry [3].

Acrolein is the simplest unsaturated aldehyde and exhibits high reactivity due to the presence of a C=C double bond conjugated with the carbonyl group. The acrolein has been used as herbicide in irrigation systems and as antimicrobial in liquid fuels, process lines, and in water recirculation systems and is a crucial intermediary in the industrial production of a wide range of compounds such as methionine, acrylic acid, acrylic acid esters, polymers, propanol, propionaldehyde, allyl alcohol, 1,3-propanediol, acrolein acetals, alkoxy-propionaldehydes, and pyridine bases [4].

The glycerol dehydration is mainly carried out in gaseous phase in the presence of an acid catalyst such as protonated or metal-promoted zeolites, mixed metallic oxides, functionalized oxides, or supported heteropolyacids [5], at atmospheric pressure and reaction temperatures between 453 and 773 K [6]. Depending on the reaction conditions and the physicochemical properties of the catalyst, acetol (1-hydroxy-2-propanone) and acetaldehyde (ethanal) may be produced by parallel dehydration routes, while small amounts of aldehydes, carboxylic acids, and/or alcohols in the range of $C_1$–$C_3$ are results of subsequent reactions of the dehydration products [7].

This chapter highlights the advances in the gas-phase catalytic dehydration of glycerine to acrolein.

## 2. Thermodynamics of the glycerol dehydration

The thermodynamic analysis of a chemical system provides valuable information for the design of chemical reactors such as the heat released or absorbed by the reaction, the behavior of simultaneous and consecutive reactions regarding the temperature, and the equilibrium concentration of each compound involved in the system at a determined temperature. In this sense, the glycerol dehydration reaction proceeds through three parallel routes as shown in **Figure 1**, from which acetol and acrolein are the main products (reactions 1 and 2), while acetaldehyde and formaldehyde may be produced in minor proportions (reaction 3) [7, 8].

The reaction enthalpies ($\Delta H_r°$) of the three parallel routes at the gas phase evidence that the production of acetol (reaction 1) is an exothermic process releasing 34 kJ·mol$^{-1}$ at 298.15 K, while the system becomes endothermic to obtain acrolein (reaction 2) and acetaldehyde (reaction 3), requiring 28.8 and 56.8 kJ·mol$^{-1}$, respectively (**Table 1**). The theoretical values of the equilibrium constants ($K_p$) indicate

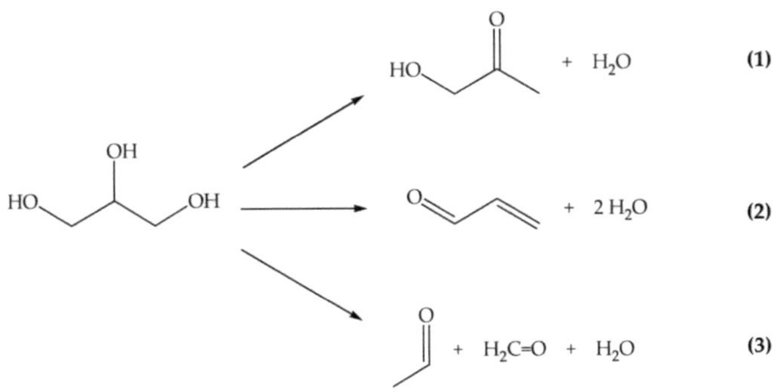


**Figure 1.**
*Parallel reactions involved in the glycerol dehydration.*

that the three reactions are thermodynamically feasible from 300 to 900 K [7]. From experimental results, Talebian et al. [9] performed calculations of equilibrium constants for the conversion of glycerol to acrolein (reaction 2) between 553 and 613 K. The trend of the equilibrium constants (from 7.6 to 7.95) is in agreement with the direction of the theoretical estimations; however, the values are smaller than the theoretical ones. The difference may be attributed to the fact that the authors considered the effect of water as solvent besides that the experimental system did not reach the chemical equilibrium, resulting in glycerol conversions smaller than the theoretical and concentrations of reactants and products that lead to different values of the thermodynamic equilibrium constant [7].

Presented in **Figure 2**, the equilibrium molar fractions ($y_i$) of each compound indicate that production of acetol prevails at mild temperatures, mainly from 300 to 480 K, attaining $y_{acetol}$ = 0.50–0.47 as its highest concentration between 300 and 400 K, while its molar fraction decreases approximately 97% from 400 to 600 K.

Contrary, the acrolein concentration increases along the reaction temperature range reaching its maximum and staying around at $y_{acrolein}$ = 0.31 between 600 and 800 K. For reaction 3, below 500 K, the degree of advancement estimated is neglectable, increasing and remaining between 500 and 800 K, which results in low molar fractions of formaldehyde and acetaldehyde, reaching a maximum value of $y_i$ = 0.034 for each product at 900 K.

| Reaction | $\Delta H_r^o$ (kJ·mol$^{-1}$) | ln ($K_p$) | | | | | | | |
|---|---|---|---|---|---|---|---|---|---|
| | | 298 K | 300 K | 400 K | 500 K | 600 K | 700 K | 800 K | 900 K |
| 1 | −33.99 | 29.39 | 29.30 | 25.91 | 23.85 | 22.43 | 21.36 | 20.53 | 19.86 |
| 2 | 28.84 | 19.11 | 19.18 | 22.15 | 23.96 | 25.09 | 25.82 | 26.28 | 26.57 |
| 3 | 56.77 | 6.93 | 7.07 | 12.77 | 16.11 | 18.24 | 19.68 | 20.71 | 21.48 |

**Table 1.**
*Standard enthalpies and equilibrium constants of glycerol dehydration reactions.*

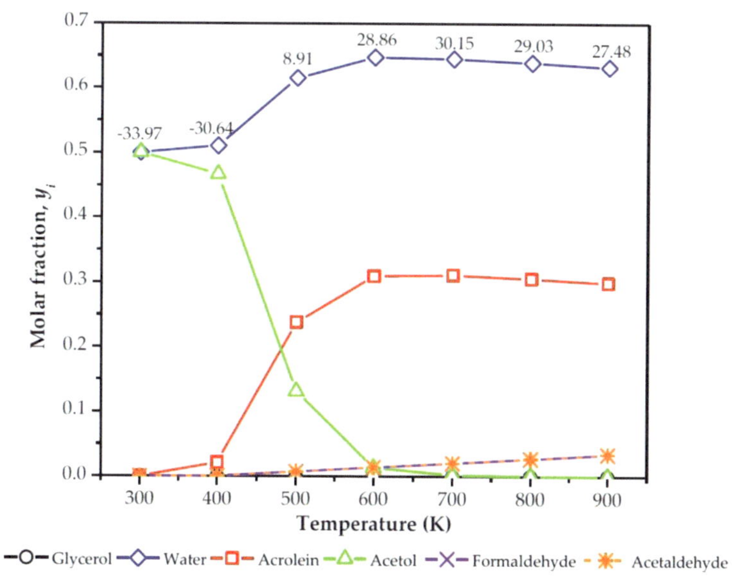


**Figure 2.**
*Equilibrium molar fractions of products as function of temperature of glycerol dehydration [7].*

On the other hand, as was expected, the molar fraction of water in the whole system shows a higher value than the rest of the compounds all over the temperature range over $y_{water}$ = 0.50 and increases to 0.64 simultaneously with the formation of acrolein. In this reaction two molecules of water are released per molecule of glycerol. The numerical values over the molar fraction curve of water indicate the heat of reaction (in kJ·mol$^{-1}$) of the overall system after an enthalpy balance, pondering the degree of advancement of each independent reaction [7].

## 3. Reactors for the glycerine dehydration in gaseous phase

Performing of the gas-phase catalytic dehydration of glycerine is usually accomplished in continuous fixed-bed and fluidized-bed reactors. These types of reactors are described in the following.

### 3.1 Fixed-bed reactors

As shown in **Figure 3(a)**, the fixed-bed reactor consists mainly on a steel alloy tube provided with an inner mesh on which the catalyst particles are deposited occupying the internal volume. A distributor tray is placed below the reactor entrance, to offer a uniform feedstock flow, as well as a layer of a nonporous and inert material such as fused ceramic on top of the catalytic bed [10]. For the catalytic dehydration of glycerine, the reactor is heated usually between 523 and 603 K. Moreover, an aqueous glycerol solution is preheated in a preheating zone at a temperature enough to vaporize the feedstock, between 473 and 533 K depending on the concentration of reactant required in the feed, and is carried by a pure inert gas flow, usually nitrogen ($N_2$), or in mixture with reactive gases like hydrogen ($H_2$) or oxygen ($O_2$) to diminish the catalyst deactivation [7, 11, 12].

The gaseous mixture of glycerine, water, and the carrier gas is continuously fed downward the reactor in nearly plug flow at a known molar or volumetric flow, regarding the reactant or the carrier gas, respectively. The output stream from the reactor may consist of a mixture of the carrier gas, water, unconverted glycerine, acrolein, and condensable and noncondensable by-products. The condensable compounds may be separated and purified by distillation, while the noncondensable products may be treated in absorption units [13].

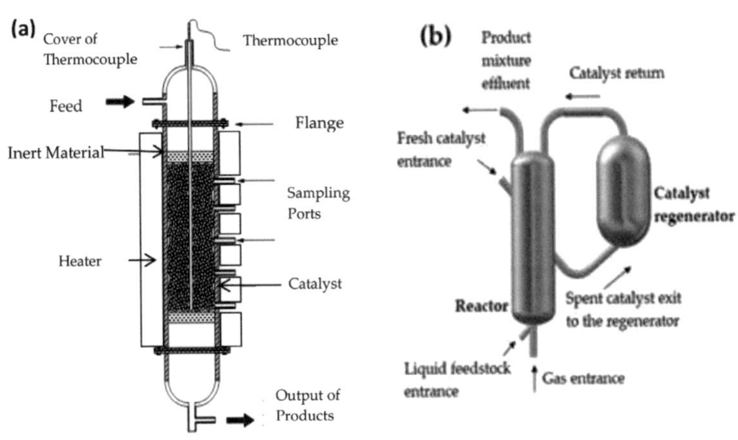


**Figure 3.**
*Schematic diagrams of reactors used in the gas-phase catalytic dehydration of glycerol to acrolein: (a) fixed-bed reactor and (b) fluidized-bed reactor.*

One of the first processes to convert glycerol into acrolein in gaseous phase using a fixed-bed reactor was patented by Schwenk et al. [14]. The authors reported the use of tubes to contain and heat bulk of supported phosphates through which pure or water-diluted glycerol vapors were passed at temperatures between 573 and 873 K. Glycerine was converted to acrolein with yields between 75 and 80% depending on the reactant concentration in the feedstock. Similarly, the patent of Neher et al. [15] reported the use of $\alpha$-$Al_3O_2$ spheres impregnated with phosphoric acid deposited in a 15-mm diameter steel tube to convert vaporized aqueous glycerol solutions to acrolein at 573 K, resulting in acrolein yields between 75 and 65% depending on the glycerol concentration in the feedstock. It is noteworthy that the catalytic activity was maintained after 60 h of operation.

### 3.2 Fluidized-bed reactors

The fluidized-bed reaction systems consist of two coupled units: the reactor itself and the catalyst regenerator as presented in **Figure 3(b)**. In the reactor, a bed of solid catalyst (with particle sizes between 7.5 and 130 μm) is initially deposited on a screen. Subsequently, a fluid (a mixture of the feedstock and a carrier gas) is fed at the bottom of the vessel passing through the catalyst at a velocity high enough to suspend and distribute the solid particles along the reactor, causing the catalyst to behave as a fluid. This process is known as fluidization. When the steady state has been reached, the catalyst is continuously fed at the top of the reactor and moved downward against the fluid stream to be removed from the fluidized bed subsequently. Once discharged from the reactor, the spent catalyst is sent directly to the regenerator where the coke is burned off with air at temperatures between 823 and 925 K. The regenerated catalyst is promptly sent back to the reactor providing the necessary heat for performing the reaction. The rate of circulation of the solids is dictated by the heat balance and the catalyst activity [16, 17].

Corma et al. [18] carried out the catalytic dehydration of glycerol in a fluidized-bed reactor in the presence of a ZSM-5-based catalyst, finding that the best operation conditions were 623 K, a catalyst/feed ratio of 11.5, residence time equal to 0.9 s, weigh hourly space velocity (WHSV) of 335 $h^{-1}$, and a concentration of 20 wt % of glycerol in the aqueous feedstock, reaching 100% of conversion and 62.1% of acrolein yield. The authors also compared the performance of this system against a fixed-bed reactor at the operating conditions. While the glycerol conversions and the product distributions were quite similar, the main difference between both processes was the higher amount of coke deposited on the catalyst used in the fixed-bed reactor (1%) than that deposited during the fluidized-bed operation (0.2%).

In other studies [19], the catalytic dehydration of a 28 wt % aqueous glycerol solution was performed at 553 K using phosphotungstic acid supported on titania ($H_3PW_{12}O_{40}/TiO_2$) as catalyst in a fluidized-bed reactor of 52 mm in height and 8 mm in internal diameter. The authors used a mixture of argon and oxygen to fluidize 1.5 g of catalyst and determined that the minimum velocity of fluidization was 1.4 $cm{\cdot}s^{-1}$; however, the catalytic tests were carried out at a velocity three times higher than this value. Under these conditions, the glycerol conversion was complete, and the acrolein yield reached 48.3%. It was found that as much as 85% of the glycerol was converted to coke in the first hour and less than 20% to acrolein. However, the acrolein selectivity increased and the coke selectivity decreased with time-on-stream (TOS).

### 3.3 Process variables

There are three process variables reported in the literature to be the most important for the catalytic dehydration of glycerine: the composition of the aqueous glycerol solution, the reaction temperature, and the space velocity. In the next sections, the effects of these variables on the catalytic dehydration of glycerol are presented.

### *3.3.1 Composition of the aqueous glycerol solution*

Since pure glycerol is highly viscous (1.5 Pa·s at 293 K) and presents a very low vapor pressure (0.05 MPa at 533.6 K) [20, 21], the use of aqueous solutions has been a strategy to overcome these drawbacks allowing the vaporization of glycerol and its use as feedstock in catalytic processes. However, the composition of the glycerine solution affects the performance of the reaction. **Figure 4** presents the results of glycerol conversion and product yields regarding the concentration of glycerol in the feedstock when using phosphotungstic acid supported on niobium pentoxide ($H_3PW_{12}O_{40}/Nb_2O_5$) as catalyst [22]. The conversion of glycerol declined from 99.8 to 94%, while the acrolein yield decreased from 91.8 to 67.7% with the increment in glycerol concentration from 10 to 40%. Similar results were observed for acetol, while for acetaldehyde there was not a clear trend. It is important to notice the enhancement in the yield of by-products (allyl alcohol, acetic acid, and unknown compounds) with the increase of glycerol in the feedstock, indicating the occurrence of side reactions. The use of other catalysts such as H-ZSM-5, H-β, H-ferrierite, silica-alumina mixtures, and supported heteropolyacids gave similar behaviors of the glycerol conversion and acrolein yield with the increase of glycerol concentration [23–26].

These results suggest that at low glycerol concentrations (large amounts of water), the water molecules may modulate side reactions of glycerol and acrolein such as etherification, oxidation, hydrogenolysis, condensation, and polymerization, thus enhancing the acrolein selectivity [23, 27]. On the contrary, with high glycerol concentrations, the diminishment in conversion and acrolein yield is attributed to the decline of the dehydration activity caused by the decrease of available active sites on the catalyst surface by glycerol condensation, promoting side

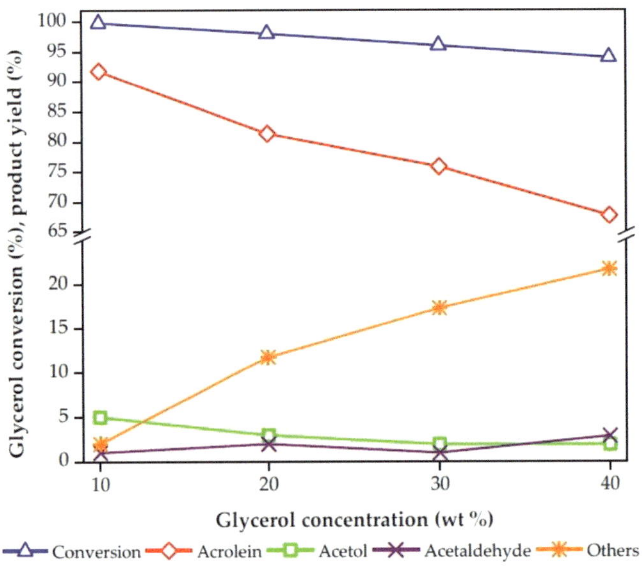


**Figure 4.**
*Effect of the glycerol concentration in the feedstock on the glycerol conversion and product yield. Data from [22].*

reactions and carbon deposition [27]. Consequently, the catalyst stability with the time-on-stream (TOS) is adversely affected when increasing glycerol content in the feed. **Table 2** summarizes this behavior, considering the effect of the water content (from 15.7 to 91.7 mol %) on the glycerol dehydration over H-ZSM-5 (150) with time-on-stream [23].

### *3.3.2 Reaction temperature*

The reactor temperature determines the products present in the glycerine dehydration reaction mixture, and according to thermodynamics, the acrolein production would be predominant from 480 K reaching its maximum at 600 K [7]. Experimentally, the increase in reaction temperature increases the glycerine conversion and therefore the acrolein yield.

**Figure 5** presents the influence of temperature on the glycerol conversion and acrolein yield for the gas-phase reaction over catalysts of 20 wt % of phosphomolybdic acid ($H_3PMo_{12}O_{40}$, HPMo), phosphotungstic acid ($H_3PW_{12}O_{40}$, HPW), and silicotungstic acid ($H_4SiW_{12}O_{40}$, HSiW) supported on commercial alumina ($Al_2O_3$, A5) in a fixed-bed reactor [28]. Above 548 K, the acrolein yield declined because the decomposition reaction toward acetaldehyde and formaldehyde is favored at high temperatures; however, the temperature at which this reaction begins to be prominent also depends on the acidity of the catalyst employed, varying from 548 to 598 K.

**Table 3** shows the effect of reaction temperature, between 553 and 593 K, and TOS on the glycerine dehydration in the presence of MCM-22 (molar ratio $SiO_2/Al_2O_3$ = 30) as catalyst [29]. As previously stated, at initial stages of the process, the glycerol conversion enhances with the temperature increase. However, severe catalyst deactivation with TOS occurs at higher temperatures. An improvement of the acrolein selectivity was also observed with the rise of temperature at initial activities, maintaining the trends along the TOS and resulting in a higher acrolein yield at 593 K even after 10 h. Similar behavior has been reported for the glycerol dehydration performed over several catalysts such as H-ZSM-5 (150), H-β (25) and H-ferrierite (55), La-$NH_4$-modified H-β (13) zeolite, and aluminosilicophosphate nanospheres (ASPN-40) [23, 24, 30, 31]. The influence of the reaction temperature on the catalyst deactivation is related to coking of the catalyst as a result of subsequent reactions between acrolein, acetol, acetaldehyde, and glycerol. At low temperature, the compounds involved in coking are glycerol and acrolein oligomers and aldol condensation products, while the increment in temperature may promote more secondary reactions of the dehydration products resulting in the formation of unsaturated, heterocyclic, and aromatic compounds of high molecular weight [27].

| Water content (mol %) | Glycerol conversion (%) | | | Acrolein yield (%) | | |
|---|---|---|---|---|---|---|
| | 2 h | 6 h | 12 h | 2 h | 6 h | 12 h |
| 15.7 | 68 | 27 | 19 | 10 | 6 | 3 |
| 51.9 | 66 | 27 | 18 | 24 | 11 | 8 |
| 76.3 | 75 | 38 | 28 | 49 | 22 | 12 |
| 91.7 | 71 | 41 | 29 | 53 | 35 | 26 |

**Table 2.**
*Effect of the water content in the feedstock on the glycerol conversion and the acrolein yield with time-on-stream. Data from [24].*

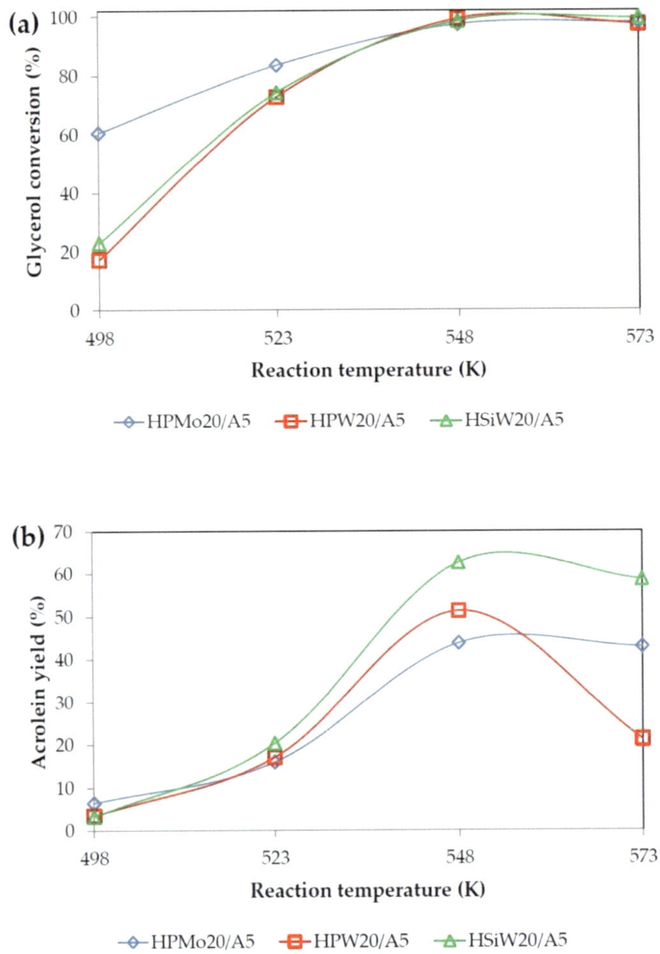


**Figure 5.**
*Effect of the reaction temperature on (a) the glycerol conversion and (b) the acrolein yield. Data from [28].*

| Temperature (K) | Glycerol conversion (%) | | | Acrolein selectivity (%) | | |
|---|---|---|---|---|---|---|
| | 1 h | 5 h | 10 h | 1 h | 5 h | 10 h |
| 553 | 80 | 44 | 33 | 22 | 15 | 8 |
| 573 | 85 | 46 | 9 | 49 | 28 | 30 |
| 593 | 100 | 48 | 22 | 54 | 42 | 22 |

**Table 3.**
*Effect of the reaction temperature on the glycerol conversion and acrolein selectivity with time-on-stream. Data from [29].*

### *3.3.3 Space velocity*

When working with continuous reactors, the space velocity is useful to relate the feed rate to the amount of catalyst. The feed rate may be expressed as the volumetric flow rate of liquid ($Q_l$), the total gas volumetric flow ($Q_g$, involving reactive and inert species), or the mass flow rate of reactant ($\dot{m}_r$), while the catalyst amount may be the volume ($V_{cat}$) or the weight of catalyst ($W_{cat}$) loaded into the reactor. The resulting terms are known as liquid hourly space velocity (LHSV), gas hourly space velocity (GHSV), and weight hourly space velocity (WHSV) which have units of reciprocal time and are defined in Eqs. 4–6. Care should be taken concerning the choice of the reference conditions, since the three ways of expressing space velocity find extensive use.

$$LHSV = \frac{Q_l}{V_{cat}} \tag{4}$$

$$GHSV = \frac{Q_g}{V_{cat}} \tag{5}$$

$$WHSV = \frac{\dot{m}_r}{W_{cat}} \tag{6}$$

**Figure 6** shows the effect of the WHSV on the glycerol conversion and yield of products of the glycerine dehydration over a Pd-HPW/Zr-MCM-41 catalyst [26]. It was evidenced that the WHSV has significant influence on the catalytic activity. The glycerol conversion increased from 90–94% with increasing WHSV from 0.17 to 0.35 $h^{-1}$. However, a further increase in WHSV led to a decrease in glycerol conversion up to 73% at 1.04 $h^{-1}$. According to the authors, this behavior was explained by the fact that increasing space velocity implies shortening the residence time for glycerol. Regarding the acrolein yield, it also presents a maximum value of 80% at 0.35 $h^{-1}$ and decreased with the increase of WHSV because the formed acrolein may further react with unconverted glycerol. This was supported by the opposite trend shown for the yield of other products (including acetic acid, allyl alcohol, and unknown products) reaching together a maximum yield of 13.9% at 1.05 $h^{-1}$. Similar results have been reported for the reaction in the presence of $NH_4$-La-modified H-β zeolite, hierarchical mesoporous H-ZSM-5 zeolites, and phosphotungstic acid supported on Cs-modified SBA-15 [30, 32–34]. Regarding the effect of space velocity on the glycerol dehydration with TOS, no marked trend was found during 20 h periods resulting in neglectable change in the glycerol conversion and acrolein yield [34].

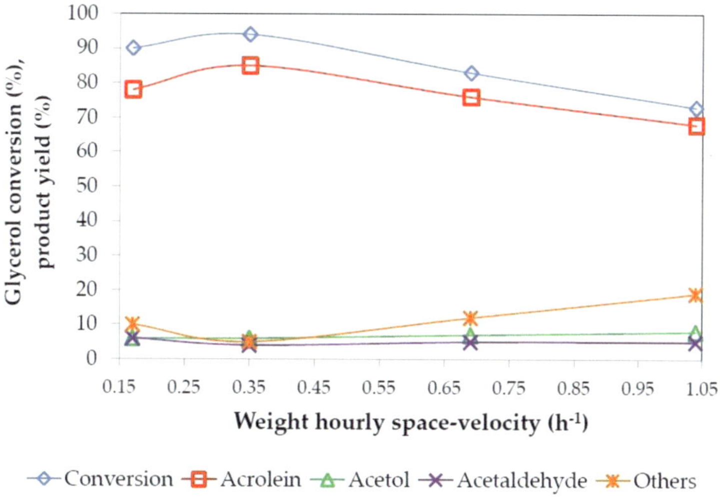


**Figure 6.**
*Effect of the weight hourly space velocity on the glycerol conversion and product selectivity. Data from [26].*

## 4. Catalysts used for the glycerine dehydration

As briefly pointed out in Section 3.1, the first attempts to perform the catalytic dehydration of glycerine were using supported mineral acids. However, the use of these catalysts involved some disadvantages, mainly the corrosive effect in pipes and vessels as well as healthy risks during their handling and rapid catalyst deactivation. On the other hand, the development of new heterogeneous catalysts during the last decades has led to an improvement of chemical processes, either in the technical, environmental, and health aspects. In this sense, during the last years, several heterogeneous acid catalysts such as protonated, metal-promoted, and hierarchical zeolites, mixed metallic oxides, functionalized oxides, and supported heteropolyacids have been evaluated to perform the catalytic dehydration to acrolein in gaseous phase. **Table 4** summarizes some relevant catalysts used in the gas-phase conversion of glycerine to acrolein, as well as the reaction conditions and their catalytic performance.

Protonated zeolites were studied by Kim et al. [23, 24] as catalysts for the glycerine dehydration in a fixed-bed reactor, taking into account several parameters such as the composition of the catalyst ($SiO_2/Al_2O_3$ molar ratio), the reaction temperature, and the amount of water in the feed. Among the tested zeolites, H-ZSM-5 (150), H-β (25), and H-ferrierite (55) showed high catalytic activities with conversions of 93.7, 95.2, and 70.9%, respectively, and acrolein yields around 53.8, 44.7 and 54.6% in the same order, at 614 K.

In other studies, Corma et al. [18] evaluated the activity of a ZSM-5-based catalyst on the conversion of glycerol/water mixtures to acrolein in a fluidized-bed reactor. The highest yield of acrolein (55–61% molar carbon yield) was obtained at 623 K with complete glycerol conversion, while the use of high temperatures (>773 K) resulted in the decrease of acrolein selectivity and the increment of several other compounds, mainly acetaldehyde, $C_1$–$C_4$ alkanes, ethylene, propylene, butenes, acetone, and organic acids.

Zeolites modified by ion-exchange have also been tested in the glycerol dehydration. Dalla et al. [30] studied the dehydration activity of the protonic (H-β) and the ammonium-lanthanum-modified beta zeolites ($NH_4$-La-β). Both zeolites reached similar initial glycerol conversions (98% and 95%, respectively, at TOS = 0.5 h) at 548 K. However, the $NH_4$-La-β zeolite was more selective toward acrolein than the protonic form, reaching 82.9% and 76.4% of acrolein yields. Additionally, the modified catalyst showed lower deactivation at 7 h of TOS than the H-β zeolite.

The activity of the Y zeolite in its protonic form (HY), with La (LaY) and Pd with La (Pd/LaY), was evaluated by Pala et al. [7] at temperatures between 473 and 573 K. The three catalysts were active in the conversion of glycerine in the temperature range. The highest conversions were 61.6, 84.1, and 93% in the order HY, LaY, and Pd/LaY at 573 K. For the three catalysts, the acrolein selectivities increased with the increase in temperature and also followed the trend LaY > HY > Pd/LaY, regarding the composition. However, the highest acrolein yields were 57.3, 75.2, and 87.6% at 573 K, for the HY, LaY, and Pd/LaY, respectively, as a result of the increase of the glycerol conversion.

The production of acrolein from glycerine in the presence of hierarchical H-ZSM-5 zeolites has proven to be feasible. Decolatti et al. [32] reported the use of the parent (Si/Al = 15) and desilicated H-ZSM-5 zeolite attaining a glycerol conversion of 62.1% and acrolein yield of 30.6% for the former at 548 K and 1 h of TOS, while the modified zeolite reached 89.6% of glycerol conversion and 72.1% of acrolein yield. Additionally, the untreated zeolite showed high deactivation resulting in 4.5% of acrolein yield after 5 h of TOS, against 58.6% reached by the desilicated zeolite. Further work of Lago et al. [33] showed that desilicated samples of

| Catalyst | Reaction conditions[a] | Performance[b] | Reactor type | Reference |
|---|---|---|---|---|
| H-ZSM-5 | 8.3 mol% glycerol, 91.7 mol% $H_2O$ in He, $F_g$ = 23.4 mmol·h$^{-1}$, T = 613 K, $W_{cat}$ = 0.30 g | $X_g$ = 93.7 $Y_{acro}$ = 53.8 | Fixed bed | [23] |
| H-Ferrierite | 8.3 mol% glycerol, 91.7 mol% $H_2O$ in He, $F_g$ = 23.4 mmol·h$^{-1}$, T = 613 K, $W_{cat}$ = 0.30 g | $X_g$ = 70.9 $Y_{acro}$ = 54.6 | Fixed bed | [24] |
| ZSM-5 mixed with clay binder | 20 wt % glycerol aqueous solution, T = 623 K, WHSV = 335 h$^{-1}$, catalyst to feed ratio = 11.5 | $X_g$ = 100 $Y_{acro}$ = 62.1 | Fluidized bed | [18] |
| $NH_4$-La-β zeolite | 20 wt % glycerol aqueous solution, T = 548 K, WHSV =0.75 h$^{-1}$, $W_{cat}$ = 0.40 g | $X_g$ = 95.0 $Y_{acro}$ = 82.9 | Fixed bed | [30] |
| Pd/LaY zeolite | 10 wt % glycerol aqueous solution, T = 573 K, GHSV = 5933 h$^{-1}$, $W_{cat}$ = 0.30 g | $X_g$ = 93.0 $Y_{acro}$ = 87.6 | Fixed bed | [7] |
| Modified H-ZSM-5 by alkaline treatment | 20 wt % glycerol aqueous solution, T = 548 K, WHSV = 0.75 h$^{-1}$, $W_{cat}$ = 0.40 g | $X_g$ = 100 $Y_{acro}$ = 74 | Fixed bed | [33] |
| $WO_3/ZrO_2$ | 30 wt % glycerol aqueous solution, T = 553 K, $V_{cat}$ = 4.5 ml, GHSV = 4400 h$^{-1}$ | $X_g$ = 100 $Y_{acro}$ = 72 | Fixed bed | [35] |
| $WO_3/TiO_2$ | 28 wt % glycerol aqueous solution, $Q_l$ = 0.5 ml·min$^{-1}$ plus a 200 ml·min$^{-1}$ Ar flow, T = 553 K, $W_{cat}$ = 100 g | $X_g$ = 100 $Y_{acro}$ = 73 | Fluidized bed | [36] |
| ZrOx-NbOx | 20 wt % glycerol aqueous solution, T = 573 K, GHSV = 1930 h$^{-1}$, $W_{cat}$ = 7.5 g | $X_g$ = 99 $Y_{acro}$ = 71.3 | Fixed bed | [37] |
| WOx-NbOx | Glycerol-water 1:5 (mol/mol) solution, T = 558 K, $W_{cat}$ = 0.20 g | $X_g$ = 98.9 $Y_{acro}$ = 74.4 | Fixed bed | [38] |
| $WO_3$-$SiO_2/ZrO_2$ | 20 wt % glycerol aqueous solution, T = 573 K, GHSV = 2900 h$^{-1}$ | $X_g$ = 100 $Y_{acro}$ = 80 | Fixed bed | [39] |
| $H_3PO_4$-MCM-41 | 20 wt % glycerol aqueous solution, T = 593 K, $W_{cat}$ = 0.30 g | $X_g$ = 97 $Y_{acro}$ = 81.5 | Fixed bed | [40] |
| SAPO-40 | 10 wt % glycerol aqueous solution, T = 593 K, WHSV = 0.85 h$^{-1}$, $W_{cat}$ = 0.30 g | $X_g$ = 100 $Y_{acro}$ = 80.6 | Fixed bed | [41] |
| $H_3PW_{12}O_{40}/TiO_2$ | 28 wt % glycerol aqueous solution, $Q_l$ = 0.5 ml·min$^{-1}$ plus an additional Ar flow, T = 553 K, $W_{cat}$ = 1.5 g | $X_g$ = 100 $Y_{acro}$ = 48.3 | Fluidized bed | [19] |
| $H_3PW_{12}O_{40}/Nb_2O_5$ | 10 wt % glycerol aqueous solution, T = 598 K, GHSV = 420 h$^{-1}$, $W_{cat}$ = 0.30 g | $X_g$ = 99.8 $Y_{acro}$ = 91.8 | Fixed bed | [22] |

| Catalyst | Reaction conditions[a] | Performance[b] | Reactor type | Reference |
|---|---|---|---|---|
| $H_3PW_{12}O_{40}$/ Cs-SBA-15 | 20 wt % glycerol aqueous solution, T = 573 K, WHSV = 0.72 $h^{-1}$, $W_{cat}$ = 0.50 g | $X_g$ = 100 $Y_{acro}$ = 86 | Fixed bed | [34] |
| $H_4SiW_{12}O_{40}/SiO_2$ | 10 wt % glycerol aqueous solution, T = 548 K, $Q_l$ = 0.028 ml·$min^{-1}$, $F_g$ = 1.8 mmol·$h^{-1}$, $W_{cat}$ = 0.30 g | $X_g$ = 98.3 $Y_{acro}$ = 86.2 | Fixed bed | [42] |
| Cs_doped $H_4SiW_{12}O_{40}/Al_2O_3$ | 20 wt % glycerol aqueous solution, T = 573 K, GHSV = 6000 $h^{-1}$ | $X_g$ = 100 $Y_{acro}$ = 88 | Fixed bed | [43] |

[a]*T = reaction temperature, WHSV = weight hourly space velocity, GHSV = gas hourly space velocity, $W_{cat}$ = weight of catalyst, $Q_l$ = liquid flow rate, $V_{cat}$ = volume of catalyst, $F_g$ = glycerol molar feed rate.*
[b]*$X_g$ = glycerol conversion (%), $Y_{acro}$ = acrolein yield (%).*

**Table 4.**
*Catalysts, reaction conditions, and performance of the catalytic dehydration of glycerol in gaseous phase.*

H-ZSM-5 zeolite resulted in an improvement of the glycerol conversion (100%) and the acrolein yield (66–74%) regarding the parent zeolite (Si/Al = 40) which reached 95% of conversion and an acrolein yield of 53% at 548 K. The desilicated zeolites maintained the glycerol conversion around 70% up to 7 h of TOS, while the acrolein yield decreased to 20% at the same time.

Catalysts of tungsten, zirconium, and niobium oxides have also shown activity in the glycerol dehydration reaction. Dalil et al. [36] investigated a catalyst of tungsten oxide supported on titania ($WO_3/TiO_2$) in a fluidized-bed reactor. Complete glycerol conversion and acrolein selectivity of 73% were reached after 6 h of TOS at 553 K. Besides the high activity of the catalyst, the authors find that the acrolein selectivity increased from 55 to 73% with the increase in TOS from 1 to 6 h, related to the increase of coke formation over the catalyst.

Lauriol-Garbay et al. [37] produced acrolein from glycerine using mixed oxides of zirconium and niobium (ZrNbO). The catalysts exhibit a selectivity to acrolein of approximately 72%, at nearly total glycerol conversion at 573 K. ZrNbO catalysts still exhibited 82% conversion efficiency after 177 h on stream, while its acrolein selectivity remains unimpaired. The catalyst calcined at 673 K achieved 98.9% of glycerol conversion and an acrolein yield of 74.4% at 558 K. The acrolein yield and the deactivation were found to be higher and slower, respectively, than those of $WO_3/ZrO_2$ and H-ZSM-5 which are typical acid catalysts [38]. In another study, Znaiguia et al. [39] got 80% of acrolein yield with complete conversion of glycerol at 573 K using a catalyst of tungstated zirconia promoted with silica (WSi/Zr). The authors confirmed that the incorporation of silicon improved the dehydration activity and the catalyst stability.

The catalytic dehydration of glycerol may also occur on oxides promoted with phosphate. Ma et al. [40] evaluated phosphorus-containing MCM-41 mesoporous molecular sieves ($H_3PO_4$-MCM-41). The catalyst with 25 mass % of supported $H_3PO_4$ resulted in 84% of acrolein selectivity with glycerol conversion of 97% at 593 K. The conversion of glycerol and selectivity to acrolein greatly depended on the calcination temperature, reaction temperature, and glycerol concentrations. Tests of the catalyst activity with TOS indicated that the HP-MCM-41 exhibited stable activity with high acrolein selectivity up to 12 h. Recently, Fernandes et al. [41] reported the use of hierarchical silicoaluminophosphate 40 (SAPO-40). When compared with the conventional SAPO-40, this catalyst showed higher acrolein selectivity (80%) at complete conversion and a catalytic lifetime up to 120 h, reaching acrolein yields between 80% and 68% during this period.

Supported heteropolyacids, mainly phosphotungstic ($H_3PW_{12}O_{40}$) and silicotungstic acid ($H_4SiW_{12}O_{40}$), and their alkali-substituted salts present high activity to convert glycerine into acrolein. Viswanadham et al. [22] studied the activity of phosphotungstic acid supported on niobium pentoxide ($H_3PW_{12}O_{40}/Nb_2O_5$) which was highly active and selective toward acrolein (glycerol conversion 98.8% and acrolein selectivity 92% at 598 K). The catalytic activity depended on the amount of heteropolyacid supported, the calcination temperature, and the reaction temperature. Tests of catalyst lifetime indicated that the solid was stable with high acrolein selectivity up to 10 h on TOS.

Liu et al. [34] used a mesoporous molecular sieve modified (SBA-15) with cesium as support for $H_3PW_{12}O_{40}$ and used the resulting solid ($H_3PW_{12}O_{40}$/Cs-SBA-15) as catalyst for the glycerol dehydration. The catalyst with 50 wt % of supported heteropolyacid reached the maximum acrolein yield (86%) and complete glycerol conversion at 573 K. Compared with the catalyst prepared with the conventional support (pure $SiO_2$), the modification of SBA-15 with Cs improved the stability of the catalyst up to 170 h of reaction, and the acrolein yield was the same as before regeneration at 773 K in air.

According to Tsukuda et al. [42], heteropolyacids supported on silica also present high activity in this reaction. The authors found that the catalytic activity depended on the type of heteropolyacid as well as the size of mesopores in the silica support. The highest activity was performed by silicotungstic acid supported on silica with mesopores of 10 nm, reaching 98.3% of glycerol conversion and 86.2% of acrolein yield at 548 K. The activity of silicotungstic acid, doped with rubidium and cesium, supported on a mixture of δ and θ $Al_2O_3$, was reported by Haider et al. [43]. The Cs-doped catalyst reached a maximum acrolein selectivity of 91% at 100% glycerol conversion for 90 h of TOS at 573 K, with a 10 wt % glycerol solution. When the glycerol concentration in the feed was increased to 20 wt %, the acrolein yield slightly decreased, and the catalyst was stable during a shorter TOS regarding the reaction with 10 wt % of glycerol in the feedstock.

The main features of these catalysts that affect the acrolein selectivity are the strength and type of the surface acid sites, which are known to promote the dehydration reactions of alcohols [44–46]. Regarding the strength of the acid sites measured in terms of the Hammett acidity (HA), the catalysts have been classified into four groups. The first group is comprised by basic catalysts with HA higher than +7 and shows no selectivity toward acrolein. Catalysts, such as zirconium oxide, with HA between −3 and + 7, belong to the second group. These solids show acrolein selectivities not greater than 30% but remain stable for 10 h on stream. Group 3 includes catalysts such as alumina impregnated with phosphoric acid, heteropolyacids supported on alumina, niobium oxide calcined at 773 K, HZSM zeolite, and pure alumina. Their HA values are between −8 and −3 and result in acrolein selectivities up to 70%; however, these catalysts show low stability and rapid deactivation. The fourth group comprehends solids with HA less than −8, such as Hβ zeolite, niobium oxide calcined at 623 K, alumina silicate, and sulfonated zirconium oxide. These catalysts are less selective to acrolein but more stable with TOS than those of group 3 [47].

Additionally, the type of acid sites present at the catalyst surface has an effect on the products' distribution. It is generally accepted that the Brønsted acidity promotes the glycerol dehydration reaction to proceed through the acrolein route (reaction 2). Some experimental studies have demonstrated the positive influence of the concentration of Brønsted acid sites on the acrolein yield, as well as the relationship of Lewis sites on the production of acetol.

In the study of Pala et al. [7], the distribution of acid sites of HY zeolite was modified by ion-exchange with La and with La and Pd. An increase in the total amount of acid sites was observed after the exchange with La cations, increasing around 1.5 and 2.1 times the concentration of Lewis and Brønsted sites in the LaY catalyst regarding the HY zeolite, at 573 K. A subsequent raise of the total acidity occurred after the impregnation of the LaY solid with Pd, leading to concentrations 2.5 and 3.5 times higher than the acidity of HY zeolite. At any temperature, the introduction of La into the HY zeolite improved the glycerol conversion, attributed to the increase of total acidity. At 573 K and GHSV = 5933 $h^{-1}$, the acrolein yield raised from 57.3% to 75.2% with the increase in the concentration of Brønsted acid sites after the modification with La. Besides, the incorporation of Pd to the LaY catalyst resulted in an acrolein yield of 87.6% at the same temperature. Since the concentration of Lewis acid sites was also increased after the ion-exchange procedures, the acetol yield followed the order Pd/LaY > HY > LaY with values of 0.07, 0.5, and 2.5%, respectively.

Kim et al. [25] reported the correlation between the acrolein and acetol yields with the concentration of Brønsted and Lewis acid sites, respectively, of a series of silica-alumina and alumina ($\eta$-$Al_2O_3$) catalysts. The acrolein yield enhanced from 3.6% to 17.2% with the increase in the concentration of Brønsted acid sites from 0 to 188 μmol $g^{-1}$, while the acetol raised from 2.2 to 5% with the change of Lewis acid sites from 28 to 192 μmol $g^{-1}$ at 588 K, WHSV = 62 $h^{-1}$, and 2 h of TOS.

Similarly, Massa et al. [48] performed the glycerol dehydration reaction over catalysts of Nb and W oxides supported on $Al_2O_3$, $SiO_2$, and $TiO_2$ at 578 K, WHSV = 0.94 $h^{-1}$, and collection of products between 1 and 3 h of TOS. The acrolein selectivity increased from 0 to 70%, presenting a sigmoidal trend regarding the increase in the concentration of Brønsted acid sites from 0 to 1 μmol $m^{-2}$. The promoting effect of Lewis acidity on the acetol production was also evidenced since the change from 0.41 to 2.95 μmol $m^{-2}$ resulted in the enhancement of the acetol selectivity from 5 to 18%, independent of the dispersed phase and the catalytic support.

## 5. Conclusions

Acrolein can be obtained from glycerine by a dehydration reaction. The main process variables in the gas phase are the reaction temperature, the concentration of glycerol in water, and the space velocity in fixed-bed reactors. A thermodynamic study of the equilibrium has been made to estimate the conversion to equilibrium as a function of temperature. The reactors are usually heated between 523 and 603 K. Some of the most active catalysts in the gas-phase reaction (yield >70%) were $NH_4$-La-β zeolite, Pd/LaY zeolite, hierarchical ZSM-5, $WO_3/ZrO_2$, $WO_3/TiO_2$, ZrOx-NbOx, WOx-NbOx, $WO_3$-$SiO_2/ZrO_2$, NbOx-WOx/$Al_2O_3$, $H_3PO_4$-MCM-41, SAPO-40, NbPSi, Pd-$H_3PW_{12}O_{40}$/Zr-MCM-41, $H_3PW_{12}O_{40}$/Cs-SBA-15, $H_3PW_{12}O_{40}/Nb_2O_5$, Cs-doped $H_4SiW_{12}O_{40}/Al_2O_3$, $H_4SiW_{12}O_{40}/TiO_2$, and $H_4SiW_{12}O_{40}/SiO_2$. In general, total conversion has been achieved at temperatures from 573 to 598 K. The catalytic process in the gas phase seems more appropriate than the liquid-phase process due to high acrolein yields and direct separation of the product effluent from the catalyst.

## Conflict of interest

The authors declare no conflict of interest.

## Author details

Israel Pala Rosas[1], Jose Luis Contreras Larios [2*], Beatriz Zeifert[1] and José Salmones Blásquez[1]

1 Escuela Superior de Ingeniería Química e Industrias Extractivas, Instituto Politécnico Nacional, Ciudad de México, México

2 CBI-Energía, Universidad Autónoma Metropolitana-Azcapotzalco, Ciudad de México, México

*Address all correspondence to: jlcl@correo.azc.uam.mx

## References


[1] Monteiro M, Kugelmeier C, Pinheiro S, Batalha M, Da Silva A. Glycerol from biodiesel production: Technological paths for sustainability. Renewable and Sustainable Energy Reviews. 2018;**88**:109-122. DOI: 10.1016/j.rser.2018.02.019

[2] Pradima J, Rajeswari M, Archna. Review on enzymatic synthesis of value added products of glycerol, a by-product derived from biodiesel production. Resource-Efficient Technologies. 2017;**3**(4):394-405. DOI: 10.1016/j.reffit.2017.02.009

[3] Liu L, Ye X, Bozell J. A comparative review of petroleum-based and bio-based acrolein production. ChemSusChem. 2012;**5**(7):1162-1180. DOI: 10.1002/cssc.201100447

[4] Etzkorn WG, editor. Kirk-Othmer Encyclopedia of Chemical Technology. 5th ed. New York, NY, USA: John Wiley & Sons, Inc.; 2007. pp. 1-29. ISBN: 978-0-471-48496-7

[5] Galadima A, Muraza O. A review on glycerol valorization to acrolein over solid acid catalysts. Journal of the Taiwan Institute of Chemical Engineers. 2016;**67**:29-44. DOI: 10.1016/j.jtice.2016.07.019

[6] Dubois JL (Arkema France). Process for manufacturing acrolein from glycerol. US 8378136 B2; 2013

[7] Pala I, Contreras JL, Salmones J, Tapia C, Zeifert B, Navarrete J, et al. Catalytic dehydration of glycerol to acrolein over a catalyst of Pd/LaY zeolite and comparison with the chemical equilibrium. Catalysts. 2017;7:73. DOI: 10.3390/catal7030073

[8] Nimlos MR, Blanksby SJ, Qian X, Himmel ME, Johnson DK. Mechanisms of glycerol dehydration. The Journal of Physical Chemistry. A. 2006;**110**:6145-6156. DOI: 10.1021/jp060597q

[9] Talebian-Kiakalaieh A, Saidina N. Kinetic modeling, thermodynamic, and mass-transfer studies of gas-phase glycerol dehydration to acrolein over supported silicotungstic acid catalyst. Industrial and Engineering Chemistry Research. 2015;**54**(33):8113-8121. DOI: 10.1021/acs.iecr.5b02172

[10] Satterfield C. Heterogeneous Catalysis in Industrial Practice. 2nd ed. New York: McGraw-Hill, Inc; 1991. ISBN: 0-07-054886-2

[11] Alhanash A, Kozhevnikova E, Kozhevnikov I. Gas-phase dehydration of glycerol to acrolein catalysed by caesium heteropoly salt. Applied Catalysis A: General. 2010;**378**:11-18. DOI: 10.1016/j.apcata.2010.01.043

[12] Nadji L, Massó A, Delgado D, Issaadi R, Rodriguez-Aguado E, Rodriguez-Castellón E, et al. Gas phase dehydration of glycerol to acrolein over $WO_3$-based catalysts prepared by non-hydrolytic sol–gel synthesis. RSC Advances. 2018;**8**:13344-13352. DOI: 10.1039/C8RA01575A

[13] Castañeda Y. Process design, simulation and optimization of acrolein production from bioglycerol [thesis]. Medellín, Colombia: Department of Process Engineering, Engineering School, EAFIT University; 2016

[14] Schwenk E, Gehrke M, Aichner F (Schering-Kahlbaum A.-G.). Production of acrolein. US 1916743; 1933

[15] Neher A (Degussa Aktiengesellschaft). Process for the production of acrolein. US 5387720; 1995

[16] Theodore L. Chemical Reactor Analysis and Applications for the

Practicing Engineer. Hoboken, NJ: John Wiley & Sons, Inc.; 2012. DOI: 10.1002/9781118158630

[17] Froment G, Bischoff K, De Wilde J. Chemical Reactor Analysis and Design. 3rd ed. Hoboken, NJ: John Wiley & Sons, Inc.; 2011. ISBN: 978-0-470-56541-4

[18] Corma A, Huber G, Sauvanaud L, O'Connor P. Biomass to chemicals: Catalytic conversion of glycerol/water mixtures into acrolein, reaction network. Journal of Catalysis. 2008;**257**:163-171. DOI: 10.1016/j.jcat.2008.04.016

[19] Dalil M, Edake M, Sudeau C, Dubois JL, Patience G. Coke promoters improve acrolein selectivity in the gas-phase dehydration of glycerol to acrolein. Applied Catalysis A: General. 2016;**522**:80-89. DOI: 10.1016/j.apcata.2016.04.022

[20] Takamura K, Fischer H, Morrow N. Physical properties of aqueous glycerol solutions. Journal of Petroleum Science and Engineering. 2012;**98-99**:50-60. DOI: 10.1016/j.petrol.2012.09.003

[21] Richardson A. Determinations of vapour-pressures of alcohols and organic acids, and the relations existing between the vapour-pressures of the alcohols and organic acids. Journal of the Chemical Society, Transactions. 1886;**49**:761. DOI: 10.1039/ct8864900761

[22] Viswanadham B, Pavankumar V, Chary K. Vapor phase dehydration of glycerol to acrolein over phosphotungstic acid catalyst supported on niobia. Catalysis Letters. 2014;**144**(4):744-755. DOI: 10.1007/s10562-014-1204-x

[23] Kim Y, Jung KD, Park ED. Gas-phase dehydration of glycerol over ZSM-5 catalysts. Microporous and Mesoporous Materials. 2010;**131**:28-36. DOI: 10.1016/j.micromeso.2009.11.037

[24] Kim Y, Jung KD, Park ED. A comparative study for gas-phase dehydration of glycerol over H-zeolites. Applied Catalysis A: General. 2011;**393**:275-287. DOI: 10.1016/j.apcata.2010.12.007

[25] Kim Y, Jung KD, Park ED. Gas-phase dehydration of glycerol over silica–alumina catalysts. Applied Catalysis B: Environmental. 2011;**107**:177-187. DOI: 10.1016/j.apcatb.2011.07.011

[26] Ma T, Yun Z, Xu W, Chen L, Li L, Ding J, et al. Pd-$H_3PW_{12}O_{40}$/Zr-MCM-41: An efficient catalyst for the sustainable dehydration of glycerol to acrolein. Chemical Engineering Journal. 2016;**294**:343-352. DOI: 10.1016/j.cej.2016.02.091

[27] Jiang X, Zhou C, Tesser R, Di Serio M, Tong D, Zhang J. Coking of catalysts in catalytic glycerol dehydration to acrolein. Industrial and Engineering Chemistry Research. 2018;**57**(32):10736-10753. DOI: 10.1021/acs.iecr.8b01776

[28] Atia H, Armbuster U, Martin A. Dehydration of glycerol in gas phase using heteropolyacid catalysts as active compounds. Journal of Catalysis. 2008;**258**:71-82. DOI: 10.1016/j.jcat.2008.05.027

[29] Carriço C, Cruz F, Santos M, Pastore H, Andrade H, Mascarenhas A. Efficiency of zeolite MCM-22 with different $SiO_2/Al_2O_3$ molar ratios in gas phase glycerol dehydration to acrolein. Microporous and Mesoporous Materials. 2013;**181**:74-82. DOI: 10.1016/j.micromeso.2013.07.020

[30] Dalla BO, Peralta MA, Querini CA. Gas phase dehydration of glycerol over, lanthanum-modified beta-zeolite. Applied Catalysis A: General. 2014;**472**:53-63. DOI: 10.1016/j.apcata.2013.12.011

[31] Choi Y, Park H, Yun Y, Yi J. Effects of catalyst pore structure and acid properties on the dehydration of glycerol. ChemSusChem. 2014;**8**:974-979. DOI: 10.1002/cssc.201402925

[32] Decolatti HP, Dalla BO, Querini CA. Dehydration of glycerol to acrolein using H-ZSM5 zeolite modified by alkali treatment with NaOH. Microporous and Mesoporous Materials. 2015;**204**: 180-189. DOI: 10.1016/j.micromeso. 2014.11.014

[33] Lago CD, Decolatti HP, Tonutti L, Dalla BO, Querini CA. Gas phase glycerol dehydration over H-ZSM-5 zeolite modified by alkaline treatment with $Na_2CO_3$. Journal of Catalysis. 2018;**366**:16-27. DOI: 10.1016/ j.jcat.2018.07.036

[34] Liu R, Wang T, Jin Y. Catalytic dehydration of glycerol to acrolein over HPW supported on $Cs^+$ modified SBA-15. Catalysis Today. 2014;**233**:127-132. DOI: 10.1016/j.cattod.2013.09.062

[35] Stošić D, Bennici S, Couturier JL, Dubois JL, Auroux A. Influence of surface acid–base properties of zirconia and titania based catalysts on the product selectivity in gas phase dehydration of glycerol. Catalysis Communications. 2012;**17**:23-28. DOI: 10.1016/j.catcom.2011.10.004

[36] Dalil M, Carnevali D, Dubois JL, Patience G. Transient acrolein selectivity and carbon deposition study of glycerol dehydration over WO3/TiO2 catalyst. Chemical Engineering Journal. 2015;**270**:557-563. DOI: 10.1016/j. cej.2015.02.058

[37] Lauriol-Garbay P, Millet JMM, Loridant S, Bellière-Baca V, Rey P. New efficient and long-life catalyst for gas-phase glycerol dehydration to acrolein. Journal of Catalysis. 2011;**280**:68-76. DOI: 10.1016/j.jcat.2011.03.005

[38] Omata K, Izumi S, Murayama T, Ueda W. Hydrothermal synthesis of W–Nb complex metal oxides and their application to catalytic dehydration of glycerol to acrolein. Catalysis Today. 2013;**201**:7-11. DOI: 10.1016/j. cattod.2012.06.004

[39] Znaiguia R, Brandhorst L, Christin N, Bellière V, Rey P, Millet JM, et al. Toward longer life catalysts for dehydration of glycerol to acrolein. Microporous and Mesoporous Materials. 2014;**196**:97-103. DOI: 10.1016/ j.micromeso.2014.04.053

[40] Ma T, Ding J, Shao R, Yun Z. Catalytic conversion of glycerol to acrolein over MCM-41 by the grafting of phosphorus species. Canadian Journal of Chemical Engineering. 2016;**94**:924-930. DOI: 10.1002/cjce.22457

[41] Fernandes A, Ribeiro M, Lourenço J. Gas-phase dehydration of glycerol over hierarchical silicoaluminophosphate SAPO-40. Catalysis Communications. 2017;**95**:16-20. DOI: 10.1016/j.catcom. 2017.02.015

[42] Tsukuda E, Sato S, Takahashi R, Sodesawa T. Production of acrolein from glycerol over silica-supported heteropoly acids. Catalysis Communications. 2007;**8**:1349-1353. DOI: 10.1016/j.catcom.2006.12.006

[43] Haider M, Dummer N, Zhang D, Miedziak P, Davies T, Taylor S, et al. Rubidium- and caesium-doped silicotungstic acid catalysts supported on alumina for the catalytic dehydration of glycerol to acrolein. Journal of Catalysis. 2012;**286**:206-213. DOI: 10.1016/j.jcat.2011.11.004

[44] Bezoukhanova CP, Kalvachev YA. Alcohol reactivity on zeolites and molecular sieves. Catalysis reviews: Science and. Engineering. 1994;**36**:125-143. DOI: 10.1080/01614949408013922

[45] Lauront-Pernot H. Evaluation of surface acido-basic properties of inorganic-based solids by model catalytic alcohol reaction networks. Catalysis reviews: Science and. Engineering. 2006;**48**:315-361. DOI: 10.1080/01614940600816634

[46] Guisnet M, Pinard L. Characterization of acid-base catalysts through model reactions. Catalysis reviews: Science and. Engineering. 2018;**60**:337-436. DOI: 10.1080/01614940.2018.1446683

[47] Katryniok B, Paul S, Belliere-Baca V, Reye P, Dumeignil F. Glycerol dehydration to acrolein in the context of new uses of glycerol. Green Chemistry. 2010;**12**:2079-2098. DOI: 10.1039/c0gc00307g

[48] Massa M, Andersson A, Finocchio E, Busca G. Gas-phase dehydration of glycerol to acrolein over $Al_2O_3$-, $SiO_2$-, and $TiO_2$-supported Nb- and W-oxide catalysts. Journal of Catalysis. 2013;**307**:170-184. DOI: 10.1016/j.jcat.2013.07.022

Chapter 3

# Production of Solketal Using Acid Zeolites as Catalysts

*Vinicius Rossa, Gisel Chenard Díaz, Germildo Juvenal Muchave, Donato Alexandre Gomes Aranda and Sibele Berenice Castellã Pergher*

## Abstract

Commercial solketal is known as Augeo™ SL 191 and is produced by Rhodia (a member of the Solvay Group), which stands out as a slow evaporation solvent derived from glycerin which is considered a renewable source. It has low toxicity to human health and the environment. It is a good solvent for resins and polymers, replacing solvents derived from petroleum, and can be used as an additive of (bio) fuels. This work aimed to study acidy zeolites (H-BEA, H-MOR, H-MFI, and H-FER) as new heterogeneous catalysts of solketal production, through the ketalization reaction of glycerol with acetone. The catalytic activity showed H-BEA > H-MOR = H-MFI > H-FER after 180 min, in kinetics study. The major conversion was 85% for H-BEA. It was also verified that all the catalysts can be reused four times without washing or pretreatment among reactions in batch reactor. The solketal produced in this work was characterized by comparing it with its commercial standard, obtaining very similar characteristics.

**Keywords:** glycerol, zeolites, ketalization, solketal, catalysis

## 1. Introduction

The main goal of the Paris Treaty is to reduce the emission of harmful gases to the planet ($CO_x$, $NO_x$, and $SO_x$), which contribute to the increase of the greenhouse effect, leading to global warming and the increase of natural disasters, among others. For this reason, the use of biodiesel has been gaining more and more space in the world market. In addition to being a renewable fuel, it plays an excellent role for the environment, reducing emissions of harmful gases to the planet, compared to fossil fuels [1].

With the success of the global biodiesel industry, the production of glycerol also had a high growth rate. For each liter of biodiesel produced, 100 mL of crude glycerol, i.e., 10%, is obtained (**Figure 1**). Even if several industrialized products use glycerol in their formulation, the amount used in these products is not yet sufficient to meet the large production [2, 3].

The development of new technologies for the processing of the high availability of glycerol has been reported in the literature as one of the most promising goals of the present day regarding the transformation of glycerol into higher added value products. In turn, in industrial processes, this transformation of glycerol into higher added value products, such as solketal, is called glycerochemistry (**Figure 2**). The

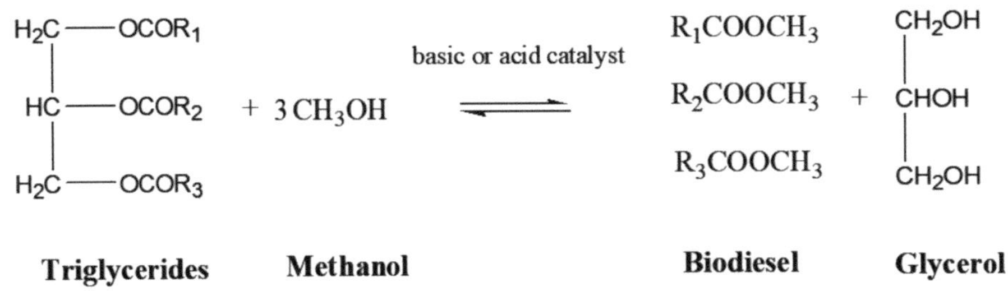


**Figure 1.**
*Reaction of biodiesel formation by triglyceride transesterification.*

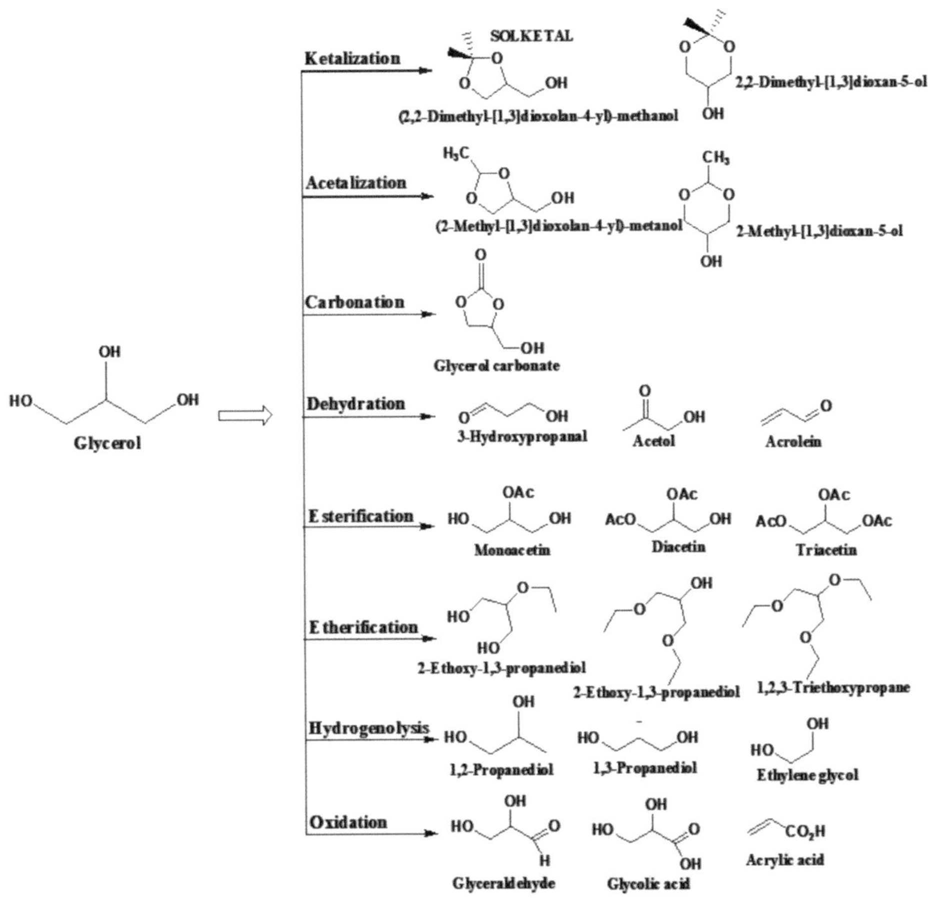


**Figure 2.**
*Transformation routes of glycerol into higher added value products.*

conversion of glycerol into other products can be accomplished by ketalization reactions, acetalization, carbonation, dehydration, esterification, etherification, hydrogenolysis, oxidation, and others [4].

In this work the focus is on the transformation of glycerol into solketal (isopropylidene glycerol or 2,2-dimethyl-1,3-dioxolan-4-yl methanol) (green solvent) through the ketalization reaction of glycerol with acetone. The reaction for solketal production is facilitated by major homogeneous and heterogeneous acid catalysts (**Figure 3**). The ketalization of glycerol with ketones generates branched oxygenates, solketal (2,2-dimethyl-[1,3] dioxan-4-yl methanol), and 2,2-dimethyl-[1,3]

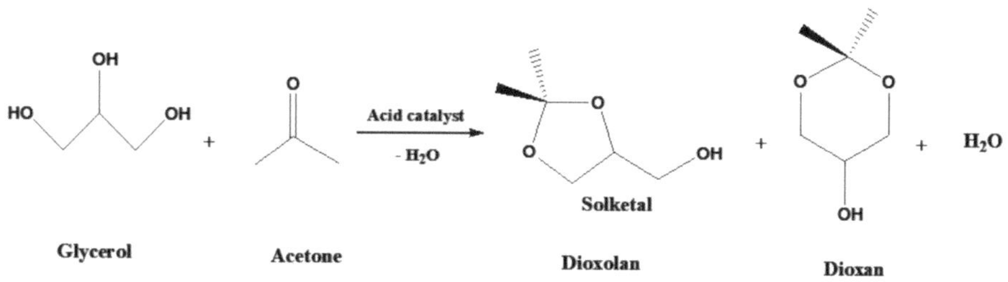


**Figure 3.**
*Solketal route production by ketalization of glycerol with acetone.*

dioxane-5-ol; however, when the reaction is carried out with acetone, the selectivity is higher for the solketal molecule, which has a five-membered ring [5].

Solketal is an excellent component for the formulation of gasoline, diesel, and biodiesel. The mixture of this compound in biofuels improves its properties, decreases viscosity, and helps to achieve the preestablished requirements for the flash point and oxidation stability of biodiesel. In addition, glycerol ketals are used as solvents, plasticizers, surfactants, disinfectants, and flavoring agents, among others. They can be used both in the pharmaceutical industry and in the food industry [5–7].

The commercial solketal is known as Augeo™ SL 191 and is produced by Rhodia (a member of the Solvay Group) and stands out as a slow evaporation solvent derived from glycerol which is considered a renewable source. Low toxicity to human health and the environment. It is a good solvent for resins and polymers, replacing solvents derived from petroleum and can be used as a (bio) fuel additive.

The synthesis of the solketal is catalyzed by p-toluenesulfonic acid (PTSA), Brönsted's acid used in homogeneous catalysis for 12 h at 100°C [8].

However, homogeneous Brönsted acid-type catalysts (hydrochloric, sulfuric, and p-toluenesulfonic acids, among others) have several disadvantages which reduce their usefulness, such as the difficulty of separation, inability to reuse, and corrosion of the reactor. The Menezes group (2013) proposed the use of homogeneous Lewis acid-type catalysts ($SnCl_2$, $SnF_2$, $Sn(OAc)_2$) that are easily recovered. The results showed that $SnCl_2$ was the most efficient and selective catalyst for solketal synthesis at room temperature. In addition to being easily recovered in the process of distillation of the reaction mixture, the catalyst can be reused for up to six times. The catalysts $SnF_2$ and Sn $(OAc)_2$ were not totally homogenized in the reaction mixture and therefore discarded. The use of $SnCl_2$ as a catalyst gave better and more economical results (at 25°C) than the conventional process, which uses PTSA as a catalyst [9].

Ferreira studied the reaction using a heteropolyacid based on phosphorus and tungsten immobilized on silica (PW-S), obtaining conversion and selectivity of 94 and 97%, respectively [10]. Silva and his group carried out tests using β-zeolite and Amberlyst™-15 polymer resin, resulting in conversions and selectivities of 90–95% and 90–95%, respectively [11]. Researchers Li et al. performed tests using MCM-41 mesoporous solid impregnated with tin (Sn-MCM-41) and USY zeolite, as catalysts, achieving conversions of 42 and 36%, respectively [12].

Khayoon [13] used activated mesoporous charcoal (AC) as a catalyst in the ketalization reaction of glycerol with acetone. The AC was impregnated with Ni and Zr, and the best result occurred when the carbon was impregnated with only 5% of metal and reached 98 and 86% selectivity, respectively. Using only active carbon as catalyst, they obtained 33% conversion and 81% selectivity. The results suggested that the improvement of these results was due to the addition of Lewis sites from nickel [13].

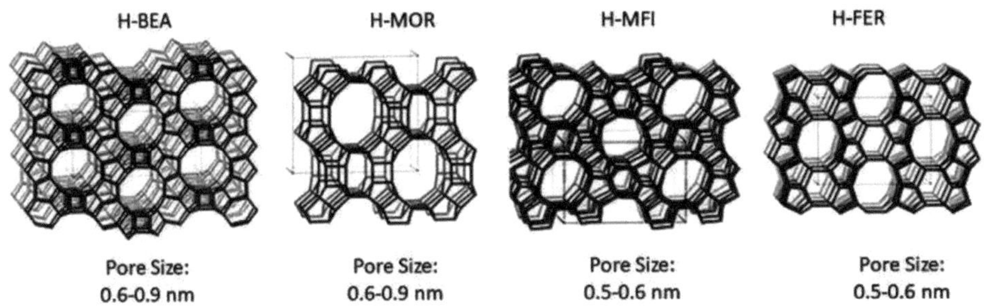


**Figure 4.**
*Structure and pore size of H-BEA, H-MOR, H-MFI, and H-FER zeolites.*

In this work, catalysts investigated for the transformation process of glycerol into solketal were acidic zeolites H-BEA, H-MOR, H-MFI, and H-FER (**Figure 4**) that can be easily recovered and reused in order to find new technological alternatives to improve the transformation routes of glycerol, a coproduct of the biodiesel industry, into products with higher added value.

## 2. Experimental

### 2.1 Reactants

The glycerol (99.5%) and acetone (99.5%) were purchased by the Brazilian industry PROQUÍMIOS.

### 2.2 Catalysts and characterizations

In this work, they used mordenite (MOR), MFI (H-ZSM-5), BEA (beta), and ferrierite (FER) acid zeolites. The zeolites were acquired in ammoniacal form, and then they were calcined at 500°C/4 h using a heating ramp of 10°C/min. After this process, the solids acquired acidic characteristics, and it is possible to use them as catalysts (H-BEA, H-MOR, H-MFI, and H-FER) in the ketalization reaction of glycerol with acetone. The catalysts were previously stored at a temperature of 100°C until the moment of reaction.

The X-ray diffraction analysis was performed to confirm the crystalline structure of the zeolites H-BEA, H-MOR, H-MFI, and H-FER before and after the calcination. The materials were characterized by X-ray diffraction in a Bruker D2 Phaser apparatus using CuKα radiation ($\lambda$ = 1.54 Å) with a Ni filter, 0.02° pitch, 0.6 mm convergent slit, 10 mA current, and voltage of 30 kV, using a Lynxeye detector.

X-ray fluorescence analysis was used to verify the chemical composition of the H-BEA, H-MOR, H-MFI, and H-FER zeolites, which made the SAR calculation (silica:alumina ratio) possible. The pastille preparation was performed using approximately 500 mg of the sample and analyzed on a Bruker XRF-S2 Ranger spectrometer.

The textural analysis was used to calculate the specific areas, volumes, and pore sizes of the H-BEA, H-MOR, H-MFI, and H-FER zeolites by the Tristar 3000 Surface Area and Porosimetry Analyzer (Micromeritics) equipment. The specific area was obtained using the BET method (Brunauer, Emmet, and Teller); the specific volume and pore diameter were obtained by the BJH method from the adsorption/desorption isotherms. The samples, after weighing, were subjected to a drying heat treatment at 300°C under a vacuum of $5 \times 10^{-3}$ torr for a period of 24 h, then cooled to room temperature, and again weighed to start the analysis at a temperature of −196°C, thus obtaining the sorption (adsorption/desorption) isotherms of $N_2$ at different partial pressures of $N_2$.

The SEM analysis was performed on a Hitachi Tabletop Microscope TM-3000 using a high-sensitivity semiconductor backscattered electron detector.

The analysis of thermoprogrammed ammonia desorption was performed to calculate the total acidic strength and to classify and to quantify the type of strength (strong/weak) of the acid sites of the zeolites. The measures TPD-$NH_3$ of ammonia were performed on a Micrometrics 2910 equipment. The procedure took place in various steps. Firstly, the catalysts are subjected to a heat treatment in order to remove impurities physically adsorbed at the acid sites of the catalyst; for this a heating rate of 10°C/min up to 550°C/30 min was used in the presence of helium gas. After, the sample was cooled to 180°C with a stream of $NH_3$ (33 mL/min) for 30 min. Then, a helium flow was passed for 90 min in order to eliminate all the ammonia adsorbed physically in the catalyst. The analysis was terminated with the thermoprogrammed desorption of the ammonia which was chemically adsorbed at the acid sites of the catalyst; at this stage a heating rate of 15°C/min was used under the helium flow (30 mL/min) at temperature range between 180 and 550°C, after remaining for 30 min at 550°C. The ammonia desorbed at different temperatures was dragged by the current of helium gas that passed through a mass spectrometer (MS), thus making it possible to calculate TPD-$NH_3$ values of the acidity sites.

### 2.3 Catalytic tests

The catalytic tests for the ketalization reaction of glycerol with acetone were conducted in a batch reactor (V = 300 mL) which was fed with 40 g of glycerol (0.43 mol), 500 rpm, 60°C, and 5% of catalyst (in relation to mass of the limiting reactant), and molar ratio glycerol:acetone was 1:4. The conversion of the glycerol and the solketal selectivity were studied. The reaction time was 180 min, and aliquots were collected to monitor reaction kinetics. All reactions of this work were performed in triplicate. The reactor has thermocouple, transducer, temperature controller external heating mantle, and agitation system. At the end of the reactions, each reaction aliquot or suspension was filtered to separate the reaction/catalyst mixture. For the partial removal of water, 2 g of anhydrous sodium sulfate was added to the obtained mixture which was again filtered. All samples were stored at 15°C until analysis by GC-FID.

### 2.4 Reuse tests

Reused reactions of the catalysts and product analysis were carried out under the same conditions as the kinetic study with a reaction time of 60 min. Each catalyst (H-BEA, H-MOR, H-MFI, and H-FER) was used in five consecutive reactions. The product was filtered, and the catalyst returned to the reactor for the next reaction without pretreatments or washes, and so on.

### 2.5 Product analysis

The products of the glycerol ketalization reaction were analyzed quantitatively by means of a Shimadzu gas chromatograph with a flame ionization detector (CG-FID), using internal standardization methodology. The column employed was Carbowax (30 × 0.25 × 0.25 μm polyethylene glycol).

The internal calibration method was applied to glycerol (99.5%) and solketal (98%) chromatographic standards using 1,4-dioxane (99.8%) as the internal standard. Moreover, a response factor was introduced to correct the GC area and obtain quantitative results.

Calculations of conversion of glycerol ($X_{Gly}$) and selectivity to solketal ($S_{Skt}$), in percentage, were obtained by the following equations:

- Conversion of glycerol ($X_{Gly\%}$):

$$X_{Gly\%} = \frac{C_{Gly_0} - C_{Gly}}{C_{Gly_0}} 100 \quad (1)$$

where $C_{Gly0}$ is the molar concentration (mol/L) of glycerol at the beginning of the reaction and $C_{Gly}$ is the molar concentration (mol/L) of glycerol at the end of the reaction.

- Selectivity ($S_{Skt\%}$):

$$S_{Skt\%} = \frac{A_{Skt}}{A_{Products}} 100 \quad (2)$$

where $A_{Skt}$ is the area of the desired product and $A_{Products}$ is the sum of the area of the solketal with the area of the unwanted products, obtained in the chromatograms.

## 2.6 Solketal characterization

At the end of the kinetic and/or catalytic tests in the GreenTec laboratory for the ketalization reaction of glycerol with acetone, the suspension remaining in the reactor composed of the glycerol-acetone/solketal-water/zeolites mixture H-BEA, H-MOR, H-MFI, and H-FER was filtered, and the liquid phase (reaction mixture) was stored in amber glass bottle. This reaction mixture was processed by rotoevaporation and fractional distillation. First, the reaction mixture was rotoevaporated to remove the remaining acetone. After the pH was adjusted between 5.5 and 6.5 with the addition of NaOH solution, 1 mol/L, to avoid the formation of other gases during the distillation process, store in a round-bottom flask containing pearls of glass, from the fractional distillation equipment (**Figure 5a**). The distillation was carried out in the Laboratory of Reactivity of Hydrocarbons, Biomass and Catalysis (LARHCO/UFRJ). Subsequently, the fractional distillation of the reaction mixture was performed (**Figure 5b**); the entire distillation process lasted for 4 h. After the distillation process, the fraction corresponding to Solketal was stored at 15°C, analyzed by FTIR, density, viscosity and water content for comparison with the analyzes of the commercial standard by the same analyzes.

The infrared spectra were obtained in a Fourier-transform infrared spectrophotometer, Nicolet model Magna-IR 760. Samples were analyzed on KBr pellets (400–4000 $cm^{-1}$). Number of scans: 16. Resolution: 4 $cm^{-1}$. The analyses were carried out in the Laboratory of Analyses of the Department of Inorganic Chemistry of IQ/UFRJ.

The specific density or mass of a sample is the ratio of the mass of a sample amount to the corresponding volume. This was determined by the Anton Paar Digital Density Meter, Model DMA 5000. The methodology for the analysis was performed according to ASTM D 4052.

Kinematic viscosity measures the flow time of a given volume of liquid flowing under the action of gravity by using a calibrated glass capillary viscometer. The method for determining this parameter was that presented in ASTM D 445.

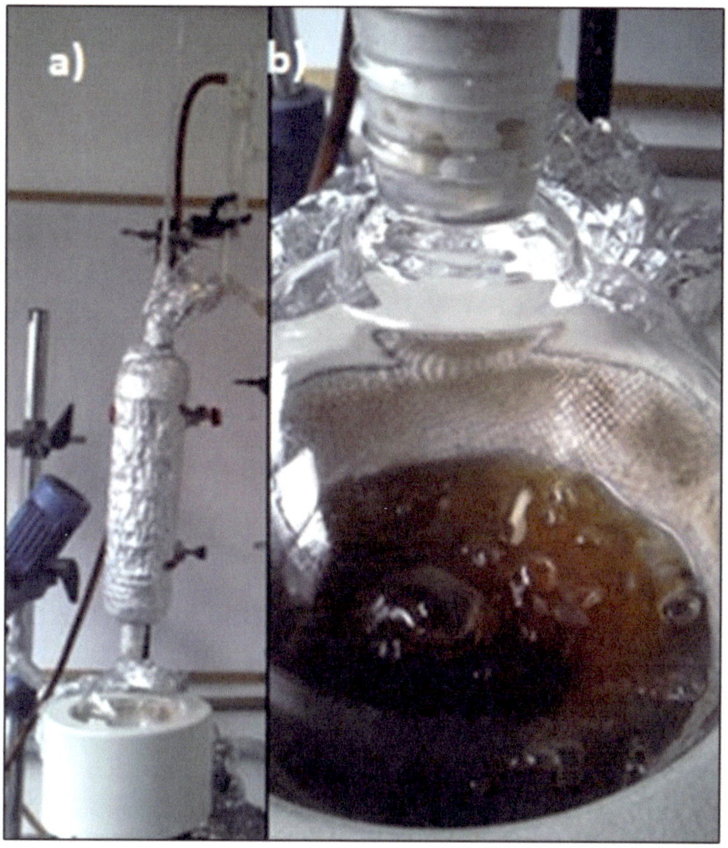


**Figure 5.**
*(a) Fractional distillation equipment and (b) reaction mixture to be distilled.*

For the water or moisture content test, the Karl Fischer coulometric method was used according to ASTM D 6304. The equipment used was the Karl Fischer METROHM titrator (model 756 KF) with the help of the Mettler XP-205 analytical balance.

## 3. Results and discussions

### 3.1 Catalysts characterization

**Figure 6** shows the X-ray diffractograms of the zeolites employed. When comparing them with those of literature [14, 15], the structure of the starting zeolites is confirmed, and that they have high crystallinity. The H-BEA, H-MOR, H-MFI, AND H-FER zeolite showed diffractograms similar to the standards of the International Zeolite Association (IZA) (www.iza-structure.org) [16].

The chemical composition of the zeolites was also studied by FRX, and the results are given in terms of oxides. The zeolites H-BEA, H-MOR, H-MFI, and H-FER are formed by oxides of silica and aluminum and also have small amounts of impurities. Through FRX analysis, it was also possible to determine the SARs ($SiO_2$/$Al_2O_3$ ratio) for each sample (**Table 1**).

Through SAR results, it is possible to have a slight notion about the Brönsted acidity of the zeolites used in this work, that is, the smaller the ratio, the greater the amount of aluminum in the structure and the greater the amount of compensation

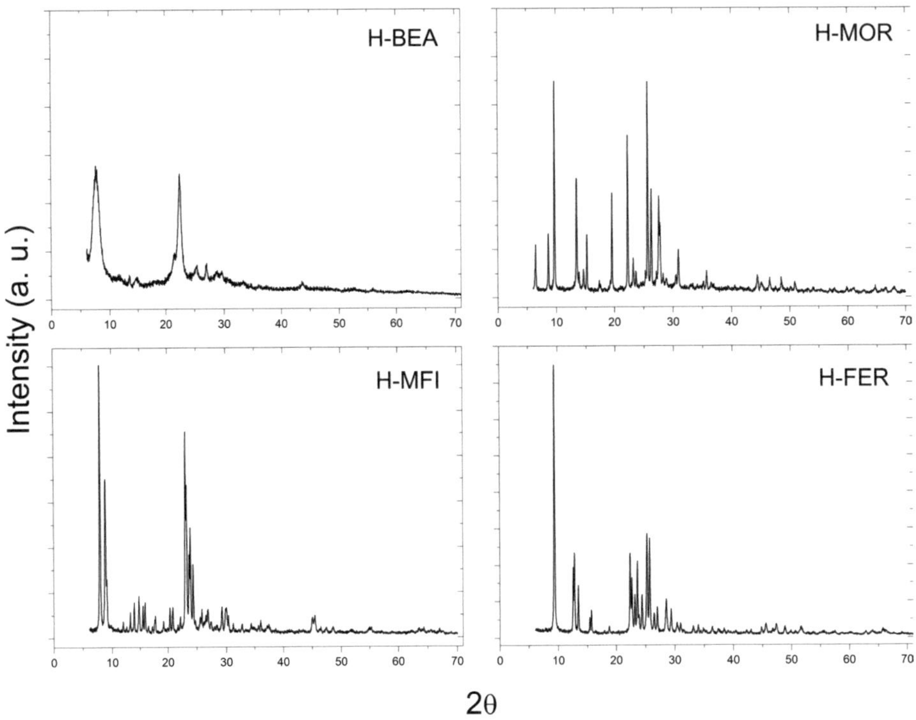


**Figure 6.**
*Diffractograms of zeolites H-BEA, H-MOR, H-MFI, and H-FER.*

| Composition (%) | Catalysts | | | |
|---|---|---|---|---|
| | H-BEA | H-MOR | H-MFI | H-FER |
| $SiO_2$ | 93.18 | 86.37 | 90.42 | 89.94 |
| $Al_2O_3$ | 5.58 | 12.38 | 7.11 | 9.15 |
| MgO | 0.50 | 0.60 | 0.50 | 0.50 |
| $Na_2O$ | 0.30 | 0.20 | 0.70 | — |
| Cl | 0.13 | 0.16 | 0.52 | 0.13 |
| $SO_3$ | 0.11 | 0.13 | 0.22 | 0.16 |
| $Fe_2O_3$ | 0.08 | 0.08 | 0.18 | 0.08 |
| $P_2O_5$ | 0.08 | — | — | — |
| $ZrO_2$ | 0.04 | — | — | — |
| $TiO_2$ | — | 0.02 | 0.21 | 0.04 |
| $ZrO_2$ | — | 0.03 | 0.05 | — |
| ZnO | — | 0.03 | — | — |
| $K_2O$ | — | — | 0.09 | — |
| SARs | 28 | 12 | 21 | 17 |

**Table 1.**
*Chemical composition and SARs of H-BEA, H-MOR, H-MFI, and H-FER zeolites by X-ray fluorescence.*

cations, providing a greater number of acidic sites in the zeolite [16, 17]. This acidity can be better studied by the analyses of TPD-NH3.

Scanning electron micrographs of the materials are shown in **Figure 7**, all on the same scale.

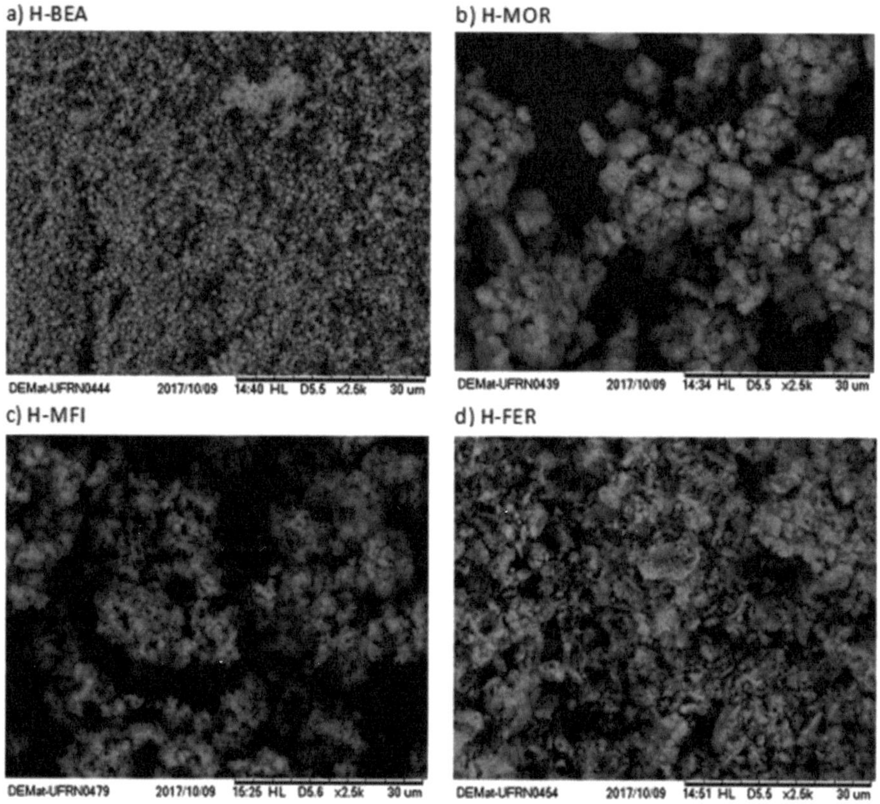


**Figure 7.**
*Micrographs of H-BEA (a), H-MOR (b), H-MFI (c), and H-FER (d) zeolites by X-ray fluorescence.*

| Characteristics | H-BEA | H-MOR | H-MFI | H-FER |
|---|---|---|---|---|
| $Area_{BET}$ [$m^2$/g] | 589 | 406 | 325 | 324 |
| $Area_{MICROPOROS}$ [$m^2$/g] | 439 | 367 | 229 | 291 |
| $Area_{EXTERNAL}$ [$m^2$/g] | 133 | 39 | 88 | 32 |
| $Volume_{TOTAL}$ [$cm^3$/g] | 0.30 | 0.21 | 0.14 | 0.17 |
| $Volume_{MICROPOROS}$ [$cm^3$/g] | 0.20 | 0.17 | 0.10 | 0.14 |
| $Volume_{BJH/DES}$ [$cm^3$/g] | 0.05 | 0.03 | 0.04 | 0.03 |
| Pore $size_{BET}$ [nm] | 2 | 2 | 2 | 2 |
| Pore $size_{BJH/DES}$ [nm] | 5 | 5 | 4 | 7 |

**Table 2.**
*Textural characteristics of H-BEA, H-MOR, H-MFI, and H-FER zeolites.*

It is observed that the H-BEA patches are smaller and appear to have a more homogeneous morphology. This is in agreement with the greater specific external area that this material has in relation to the others. The other materials have a more heterogeneous particle size distribution, and no definite shape is observed for the crystals. This may be due to the scale used in the micrograph or because they are commercial materials, which usually have a more heterogeneous morphology than those synthesized in the laboratory (on a lower-scale production).

By the adsorption analysis of $N_2$ (**Table 2**), it was observed that the materials are essentially microporous, characteristic of zeolitic materials. The H-BEA has the

| Acidity ($mmolNH_3/g_{Cat}$) | | | | |
|---|---|---|---|---|
| **Catalyst** | **Weak sites** | **Strong sites** | **Total acidity** | **F/f*** |
| H-BEA | 1.42 | 2.34 | 3.76 | 1.65 |
| H-MOR | 1.67 | 2.32 | 3.93 | 1.42 |
| H-MFI | 1.51 | 2.10 | 3.61 | 1.39 |
| H-FER | 1.60 | 1.83 | 3.43 | 1.14 |

**Ratio strong site: weak site.*

**Table 3.**
*Quantification of weak, strong, and total acidity sites.*

higher specific area due to its higher contribution of micropores and external area (due to smaller particle sizes).

Ammonia is often used as a probe molecule in acidity analyses because it has small molecular size, is stable, and possesses strong basic strength. The $NH_3$ thermoprogrammed desorption results of the H-BEA, H-MOR, H-MFI, and H-FER zeolites are shown in **Table 3**.

The zeolite H-BEA had the superior acid strength (2.34 $mmolNH_3/g_{cat}$) and bigger ratio of the strong:weak sites (1.65) than the other zeolites since this characteristic contributes to a higher catalytic activity for this reaction.

## 3.2 Kinetics and catalytic tests

The kinetic study, using the H-BEA, H-MOR, H-MFI, and H-FER catalysts, was carried out under the conditions chosen from the catalytic tests: 60° C, 550 rpm, 5% catalyst, and molar ratio glycerol:acetone 1:4. In this step, the results of conversion to glycerol and selectivity to the solketal were followed (**Figures 8** and 9).

In analyzing **Figures 8** and **9**, it is noted that the H-BEA catalyst gave the higher glycerol conversion, reaching 85% at 180 min. However, the H-FER catalyst ends

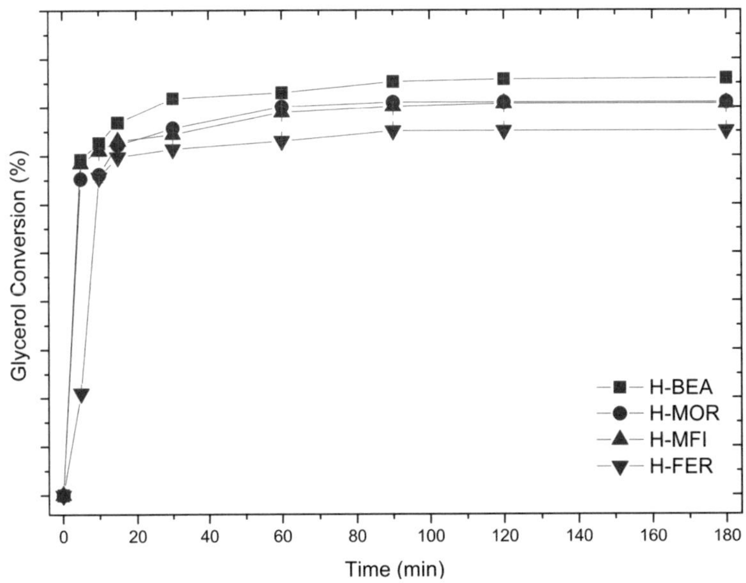


**Figure 8.**
*Glycerol conversion (%) using H-BEA, H-MOR, H-MFI, and H-FER zeolites as catalysts at 60° C, 550 rpm, 5% catalyst, and molar ratio glycerol:acetone 1:4.*

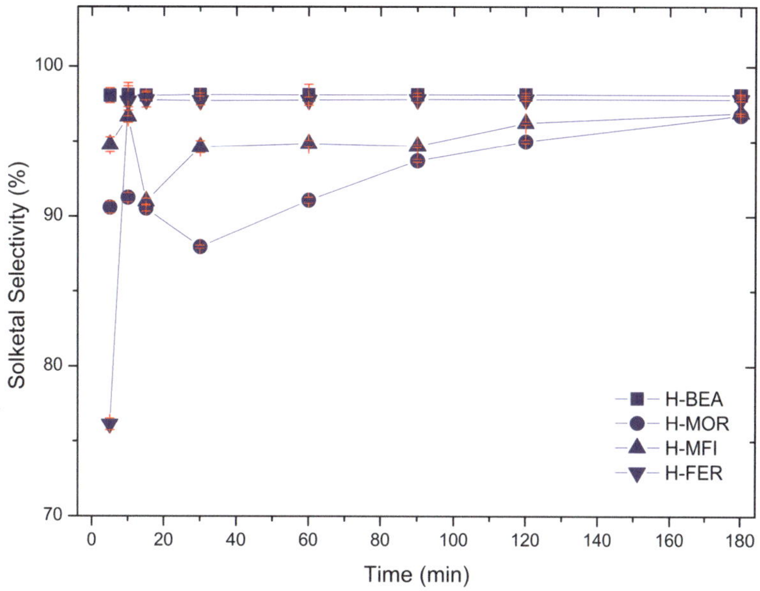


**Figure 9.**
*Solketal selectivity (%) using H-BEA, H-MOR, H-MFI, and H-FE zeolites as catalysts at 60° C, 550 rpm, 5% catalyst, and molar ratio glycerol:acetone 1:4.*

the reaction with 75% of glycerol conversion, while H-MOR and H-MFI had the same conversion, practically, at the end of the reaction, 180 min.

**Figure 8** shows that the most selective catalysts to solketal were H-BEA and H-MFI catalysts presenting values practically constant, on average 98 and 96%, respectively.

### 3.3 Catalytic activity

Many researchers have attempted to explain the activity and catalytic deactivation of the heterogeneous catalysts used in the ketalization reaction of glycerol with acetone using heterogeneous catalysts [17–25].

One factor that determines the activity of H-BEA zeolites is the size of the catalyst particle size and the presence of strong acid sites. The higher the SAR of the zeolite, the higher its acid strength and the lower the number of acid sites [17].

It is not only the acid sites of the catalyst that play an important role in the catalytic activity for this reaction but also the porosity of the catalyst [18]. Frisch (2003) concluded that the kinetic diameter of the reactants and the products of this reaction are in the range of 0.43–0.51 nm [19].

According to the literature, both Lewis acid and Brönsted acid sites are active for the ketalization reaction of glycerol with acetone [18–24].

Therefore, all the results obtained in this work can be based on the characteristics of the catalysts previously discussed in the literature. They can also help explain the data obtained in this work with zeolites H-BEA, H-MOR, H-MFI, and H-FER as catalysts. The characteristics of H-BEA contributed to it being the most active catalyst in the ketalization reaction of glycerol with acetone, when compared with the characteristics of H-FER. The zeolites H-MOR and H-MFI showed catalytic intermediate activity.

Even though acid catalysts have high activity, they can be deactivated by blocking the active sites by water molecules formed during the reaction. The higher the hydrophobicity of the catalyst, the lower the number of acid sites. However, hydrophobic groups act at the glycerol/acetone interface, reducing the interference of water molecules on the surface of the catalyst [20, 25].

Another approach, in terms of turnover frequency (TOF), the number of glycerol moles converted by moles of acidic site catalysts per hour, was performed at 60°C and 180 min, according to Eq. (3), [19]:

$$TOF = \frac{\left(\frac{NA_0 - NA}{N_{cat}}\right)}{t} \tag{3}$$

where $NA_0$ is amount of matter of glycerol at the initiation of the reaction, $NA$ is the amount of matter of glycerol in x time (x = 1, 2, and 3 h.), $N_{cat}$ is the amount of matter of the acidic site catalyst, and t is the time in hours (h). The *TOF* values at 1, 2, and 3 h are presented in **Figure 10**.

The value of TOF tends to decrease with the passage of time. This observation is explained by the loss of catalytic activity or by the reaction reaching an equilibrium for reversible reactions. TOF values were practically the same, 47 $h^{-1}$ for H-BEA, H-MFI, and H-FER catalysts. However, the H-MOR catalyst presented TOF a little inferior, 44 $h^{-1}$, when compared to the other catalysts, in all times of reaction. This means that from this perspective, all catalysts have basically the same activity.

### 3.4 Reuse testing

The reuse experiments were done for all catalysts under the same conditions used in the kinetic study. To perform these experiments, at the end of the reaction, the catalyst was only separated from the reaction solution by filtration and then reused four more times in the same manner. The reason why we chose to perform the reuse tests without the need for pretreatments (washing and calcination) for the catalysts between the reactions was to avoid the loss of time in the reuse tests and loss of material by manipulation. Because, industrially, it is not feasible to stop the

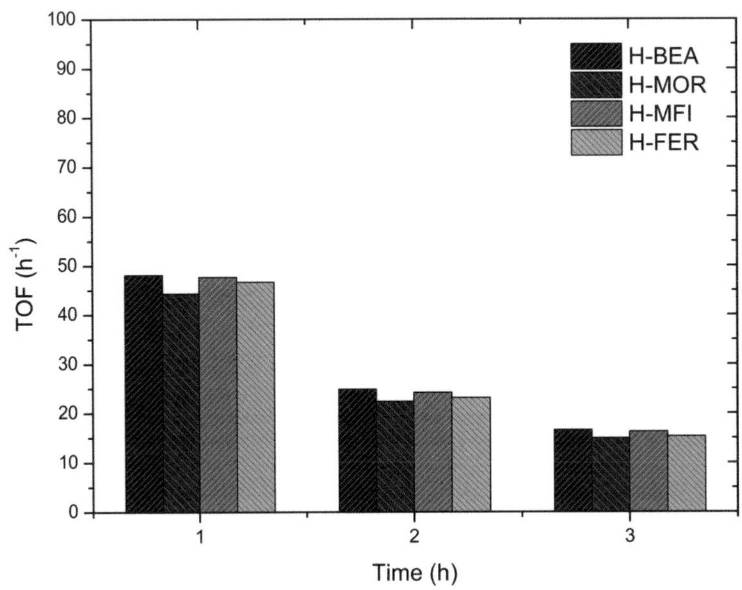


**Figure 10.**
*Results of catalytic activity by TOF using H-BEA, H-MOR, H-MFI, and H-FER zeolites as catalysts at 60°C, 550 rpm, 5% catalyst, and molar ratio glycerol:acetone 1:4, for 1, 2, and 3 h.*

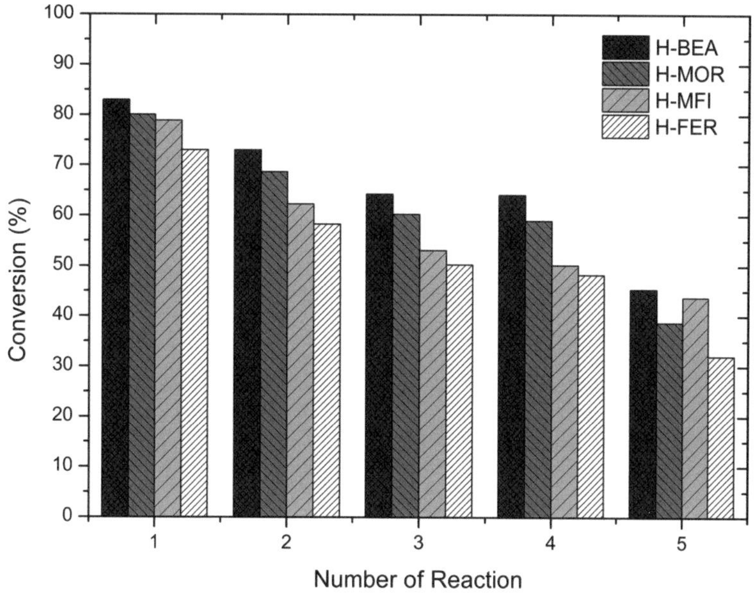


**Figure 11.**
*Solketal conversion (%) in reuse tests using H-BEA, H-MOR, H-MFI, and H-FER zeolites as catalysts at 60°C, 550 rpm, 5% catalyst, and molar ratio glycerol:acetone 1:4, 60 min.*

production to wash and calcine the catalyst every 60 min, the study was carried out in a batch reactor.

**Figure 11** shows the glycerol conversion results after each reuse experiment of the H-BEA, H-MOR, H-MFI, and H-FER catalysts.

According to **Figure 11**, the H-BEA catalyst in its first use shows an excellent conversion of glycerol to the ketalization reaction of glycerol with acetone, reaching 85% conversion of glycerol and selectivity 98%. In the fourth reuse, the conversion of glycerol reaches 55% on average. The solketal selectivity remains constant throughout the process ≈98%. For the H-MOR catalyst, the conversion of glycerol drops gradually, after each reuse, and varies between 80 and 35%, and solketal selectivity remains constant, 96%. For the H-MFI catalyst, the conversion of glycerol drops gradually, after each reuse, and varies between 80 and 45%, and selectivity to solketal varies between 95 and 90%. For the H-FER catalyst, the conversion of glycerol drops gradually, after each reuse, and varies between 75 and 30%. And its selectivity to solketal varies between 90 and 85%.

It is known that some of the water produced during the reaction and/or impurities of the reactants such as sodium residues in the glycerol together with the catalyst friction on the reactor walls have destabilized the structure of the H-BEA, H-MOR, H-MFI, and H-FER catalysts, altering their crystallinity and assisting in deactivation [4].

### 3.5 Product characterization

During the distillation process of the reaction products mixture, it was observed that when the distillation is carried out under vacuum between the temperatures 30 and 69°C, it occurs that the output of the remaining acetone and water between 70 and 120°C plus a fraction containing solketal is distilled. Glycerol is only removed when the system reaches 200°C. The yield of the distillation was 60% by mass of solketal over the initial blend (solketal-water-glycerol-traces of acetone). The solketal fraction is colorless but with a lower viscosity than glycerol.

**Figure 12.**
*Appearance of the GreenTec solketal fraction.*

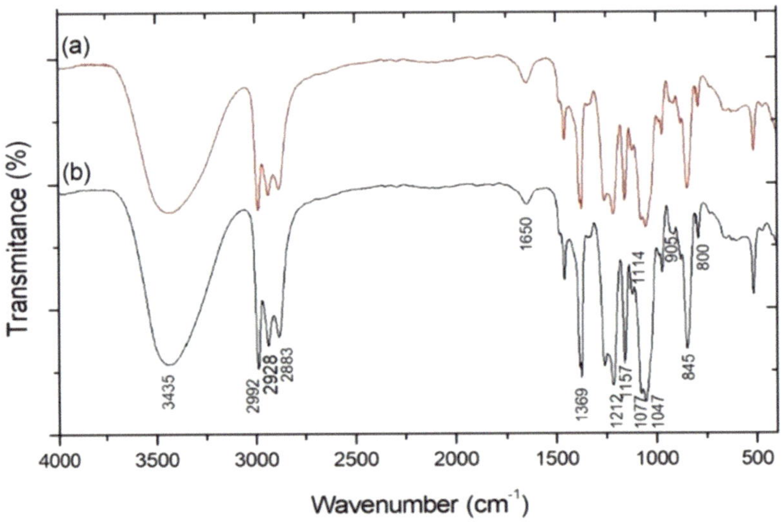


**Figure 13.**
*Infrared spectra of solketal GreenTec (a) and solketal sigma-Aldrich (b).*

**Figure 12** shows the appearance of the solketal GreenTec fraction after distillation of the initial blend.

FTIR analysis was used to confirm the presence of solketal in the distilled product and to compare it with its Sigma-Aldrich standard. The FTIR spectrum of the solketal GreenTec and solketal Sigma-Aldrich samples is shown in **Figure 13**.

**Figure 13** shows a strong band at 3435 cm$^{-1}$ which is attributed to the axial deformation of the —O —H bonds originating from the hydroxyls and hydrogen bonds between solketal molecules. The region comprising the wave numbers between 2992 and 2883 cm$^{-1}$ refers to the symmetrical axial deformation bands and asymmetric axial deformation of the —C —H bonds of the methyls.

| **Solketal** | **Density** | **Viscosity** |
|---|---|---|
| | **[g/cm³] ± DP** | **[mm²/s] ± DP** |
| Sigma-Aldrich | 1.0666 ± 0.0006 | 5.2030 ± 0.0008 |
| GreenTec | 1.0584 ± 0.0148 | 5.1650 ± 0.0123 |

**Table 4.**
*Results of density or specific gravity analysis and kinematic viscosity for solketal.*

| **Solketal** | **Water content** | | | |
|---|---|---|---|---|
| | [ppm] | [%] | [g/L] | [mol/L] |
| Sigma-Aldrich | 390 | 0.0390 | 0.3900 | 0.0217 |
| GreenTec | 610 | 0.0610 | 0.6100 | 0.0334 |

**Table 5.**
*Results of water content analysis for Solketal.*

The angular deformation of the water was attributed in the localized band appearing at 1650 $cm^{-1}$. The band located at 1369 $cm^{-1}$ refers to the movement of "umbrella" referring to the methyls of the ketone group. The bands observed at 1212 and 1077 $cm^{-1}$ refer to the —C —O bonds of the five-membered ring (dioxolanes), bands of greater importance.

The bands at 1157 and 1114 $cm^{-1}$ refer to the asymmetric vibrational motion of the —C —O —C —O —C — bonds of solketal. However, bands between 905 and 800 $cm^{-1}$ are attributed to the symmetrical vibrational movement of these same bonds. To finalize, the band located at 1047 $cm^{-1}$, the —C —C —OH bond of the alcoholic group 4 was assigned.

**Tables 4** and **5** present the results of density, viscosity, and water or moisture content for the solketal Sigma-Aldrich standard and the solketal produced by GreenTec.

When analyzing **Table 4**, it is observed that both solketal Sigma-Aldrich and solketal GreenTec present very close densities and viscosities.

**Table 5** shows that only in the analysis of humidity a significant difference between the solketal samples was noticed.

Solketal GreenTec presents 56.41% more humidity than solketal Sigma-Aldrich. To remove this moisture, anhydrous sodium sulfate may be added among other drying agents, and/or the solketal GreenTec fraction is withdrawn from 75°C.

## 4. Conclusions

Glycerol to solketal transformation is possible to carry out using zeolite acidic catalysts, such as H-BEA, H-MOR, H-MFI, and H-FER, showing a very good activity (conversion 85%) and selectivity (98%). H-BEA presented a larger area, major SAR, and a bigger ratio of the strong:weak sites than the other zeolites. This characteristic contributes to a higher catalytic activity for H-BEA catalyst. All the catalysts can be reused for four times without washing or pretreatment among reactions in batch reactor, but the best catalyst is still the H-BEA zeolite for being more active and showing constant solketal selectivity. The solketal produced in this work was characterized by comparing it with its commercial standard, obtaining very similar characteristics.

## Acknowledgements

The authors thank the following: CAPES, CNPq, TPQB/EQ/UFRJ, IQ/UFRJ, and William Xavier.

## Conflict of interest

There are no conflicts of interest in this publication.

## Appendices and nomenclature

| | |
|---|---|
| H-BEA | acid beta zeolite |
| H-MOR | acid mordenite zeolite |
| H-MFI | acid ZSM-5 zeolite |
| H-FER | acid ferrierite zeolite |
| TPD-NH3 | temperature-programmed desorption of ammonia |
| FTIR | Fourier-transform infrared spectroscopy |
| TOF | turnover frequency |

## Author details

Vinicius Rossa[1*], Gisel Chenard Díaz[2], Germildo Juvenal Muchave[2], Donato Alexandre Gomes Aranda[2] and Sibele Berenice Castellã Pergher[3]

1 Federal University of Uberlândia, Uberlândia, MG, Brazil

2 Federal University of Rio de Janeiro, Rio de Janeiro, RJ, Brazil

3 Federal University of Rio Grande do Norte, Natal, RN, Brazil

*Address all correspondence to: vinnyrossa@gmail.com

## References

[1] Pinto AG, Guarieiro LN, Resende MJC. Biodiesel: An overview. Journal of the Brazilian Chemical Society. 2005;**513**:1313-1330. DOI: 10.1590/S0103-50532005000800003

[2] Adhikari S, Fernando S, Haryanto A. Production of hydrogen by steam reforming of glycerin over alumina-supported metal catalysts. Catalysis Today. 2007;**129**:355-364. DOI: 10.1016/j.cattod.2006.09.038

[3] Sánchez EA, D'angelo MA, Comelli RA. Hydrogen production from glycerol on $Ni/Al_2O_4$ catalyst. International Journal of Hydrogen Energy. 2010;**35**:5902-5907. DOI: 10.1016/j.ijhydene.2009.12.115

[4] Rossa V. Catalytic systems for the conversion of glycerol to solketal—"Green Solvent" [thesis]. Rio de Janeiro: Federal University of Rio de Janeiro; 2017

[5] Royon D, Locatelli S, Gonzo EE. Ketalization of glycerol to solketal in supercritical acetone. Journal of Supercritical Fluids. 2011;**58**:88-92. DOI: 10.1016/j.supflu.2011.04.01

[6] Mota CJA, Silva CXX, Gonçalves VLC. Gliceroquímica: Novos produtos e processos a partir da glicerina de produção de biodiesel. Química Nova. 2009;**32**:639-648. Available from: http://www.scielo.br/pdf/qn/v32n3/a08v32n3.pdf

[7] Reddy PS, Sudarsanam P, Mallesham B, Raju G, Reddy BM. Acetalisation of glycerol with acetone over zirconia and promoted zirconia catalysts under mild reaction conditions. Journal of Industrial and Engineering Chemistry. 2011;**17**: 377-381. DOI: 10.1016/j.jiec.2011.05.008

[8] Suriyaprapadilok N, Kitiyanan B. Synthesis of solketal from glycerol and its reaction with benzyl alcohol. Energy Procedia. 2011;**9**:63-69. DOI: 10.1016/j.egypro.2011.09.008

[9] Menezes FDL, Guimaraes MDO, Silva MJD. Highly selective $SnCl_2$-catalyzed Solketal synthesis at room temperature. Industrial and Engineering Chemistry Research. 2013;**52**:16709-16713. DOI: 10.1021/ie402240j

[10] Ferreira P, Fonseca IM, Ramos AM, Vital J, Castanheiro JE. Valorisation of glycerol by condensation with acetone over silica-included heteropolyacids. Applied Catalysis B: Environmental. 2010;**98**:94-99. DOI: 10.1016/j.apcatb.2010.05.018

[11] Silva B, Figueiredo H, Santos VP, Pereira MFR, Figueiredo JL, Lewandowsk EA, et al. Reutilization of Cr-Y zeolite obtained by biosorption in the catalytic oxidation of volatile organic compounds. Journal of Hazardous Materials. 2011;**192**:545-553. DOI: 10.1016/j.jhazmat.2011.05.056

[12] Li L, Korányi TI, Sels BF, Pescarmona PP. Highly-efficient conversion of glycerol to solketal over heterogeneous Lewis acid catalysts. Green Chemistry. 2012;**14**:1611-1619. DOI: 10.1039/C2GC16619D

[13] Khayoon MS, Hameed BH. Solventless acetalization of glycerol with acetone to fuel oxygenates over Ni–Zr supported on mesoporous activated carbon catalyst. Applied Catalysis A: General. 2013;**464-465**:191-199. DOI: 10.1016/j.apcata.2013.05.035

[14] Mignoni ML, Detoni C, Pergher SBC. Estudo da síntese da zeólita ZSM-5 a partir de argilas naturais. Química Nova. 2007;**30**(1):45-48. DOI: 10.1590/S0100-40422007000100010

[15] Treacy MMJ, Higgins JB. Collection of Simulated XRD Powder Patterns for Zeolites. 5th ed. Amsterdan:

Elsevier; 2007. pp. 11-480. ISBN: 978-0-444-53067-7

[16] Sheemol VN, Tyagi B, Jasra RV. Acylation of toluene using rare earth cation exchanged zeolite β as solid acid catalyst. Journal of Molecular Catalysis A: Chemical. 2004;**215**:201-208. DOI: 10.1016/j.molcata.2004.02.002

[17] Braga A, Morgon NH. Descrições estruturais cristalinas de zeólitos. Química Nova. 2007;**30**:178-188. DOI: 10.1590/S0100-40422007000100030

[18] Gadamsetti S, Rajan NP, Rao GS, Chary KVR. Acetalization of glycerol with acetone to bio fuel additives over supported molybdenum phosphate catalysts. Journal of Molecular Catalysis A: Chemical. 2015;**410**:49-57. DOI: 10.1016/j.molcata.2015.09.006

[19] Rossa V, Pessanha Y d SP, Díaz GC, Câmara LDT, Pergher SBC, Aranda DAG. Reaction kinetic study of Solketal production from glycerol Ketalization with acetone. Industrial and Engineering Chemistry Research. 2017;**56**(2):479-488. DOI: 10.1021/acs.iecr.6b03581

[20] Sandesh S, Halgeri AB, Shanbhag GV. Utilization of renewable resources: Condensation of glycerol with acetone at room temperature catalyzed by organic-inorganic hybrid catalyst. Journal of Molecular Catalysis A: Chemical. 2015;**401**:73-80. DOI: 10.1016/j.molcata.2015.02.015

[21] Kowalska-kus J, Held A, Nowinska K. Enhancement of the catalytic activity of H-ZSM-5 zeolites for glycerol acetalization by mechanical grinding. Reaction Kinetics, Mechanisms and Catalysis. 2016;**117**:341-352. DOI: 10.1007/s11144-015-0922-4

[22] Nanda MR, Zhang Y, Yuan Z, Qin W, Ghaziaskar HS, Xu CC. Catalytic conversion of glycerol for sustainable production of solketal as a fuel additive: A review. Renewable and Sustainable Energy Reviews. 2016;**56**:1022-1031. DOI: 10.1016/j.rser.2015.12.008

[23] Stawicka K, Díaz-álvarez AE, Calvino-casilda V, Trejda M, Bañares MA, Ziolek M. The role of Brønsted and Lewis acid sites in acetalization of glycerol over modified mesoporous cellular foams. Journal of Physical Chemistry C. 2016;**120**:16699-16711. DOI: 10.1021/acs.jpcc.6b04229

[24] Venkatesha NJ, Bhat YS, Jai Prakash BS. Dealuminated BEA zeolite for selective synthesis of five-membered cyclic acetal from glycerol under ambient conditions. RSC Advances. 2016;**6**:18824-18833. DOI: 10.1039/C6RA01437B

[25] Silva CXAD, Mota CJA. The influence of impurities on the acid-catalyzed reaction of glycerol with acetone. Biomass and Bioenergy. 2011;**35**:3547-3551. DOI: 10.1021/ef9015735

Chapter 4

# Enzymatic Synthesis of Functional Structured Lipids from Glycerol and Naturally Phenolic Antioxidants

*Jun Wang, Linlin Zhu, Jinzheng Wang, Yan Hu and Shulin Chen*

## Abstract

Glycerol is a valuable by-product in biodiesel production by transesterification, hydrolysis reaction, and soap manufacturing by saponification. The conversion of glycerol into value-added products has attracted growing interest due to the dramatic growth of the biodiesel industry in recent years. Especially, phenolic structured lipids have been widely studied due to their influence on food quality, which have antioxidant properties for the lipid food preservation. Actually, they are triacylglycerols that have been modified with phenolic acids to change their positional distribution in glycerol backbone by enzymatically catalyzed reactions. Due to lipases' fatty acid selectivity and regiospecificity, lipase-catalyzed reactions have been promoted for offering the advantage of greater control over the positional distribution of fatty acids in glycerol backbone. Moreover, microreactors were applied in a wide range of enzymatic applications. Nowadays, phenolic structured lipids have attracted attention for their applications in cosmetic, pharmaceutical, and food industries, which definitely provide attributes that consumers will find valuable. Therefore, it is important that further research be conducted that will allow for better understanding and more control over the various esterification/transesterification processes and reduction in costs associated with large-scale production of the bioconversion of glycerol. The investigated approach is a promising and environmentally safe route for value-added products from glycerol.

**Keywords:** glycerol, phenolic antioxidant, phenolic structured lipids, microreactors, lipase

## 1. Introduction

Glycerol, also known as glycerin or propane-1,2,3-triol, is a chemical which has a multitude of uses in pharmaceutical, cosmetic, and food industries [1]. Currently the modifications of glycerol such as mono-, di-, and triglycerides are representing valuable products, as they have numerous applications such as modifying agents in food and pharmaceutical industries [2]. Especially, the triglycerides, which are also called structured lipids, have been widely concerned by researchers. Structured lipids are tailor-made fats and oils with special metabolic methods by incorporating new fatty acids or changing the position of existing fatty acids on the glycerol backbone. By adding a special functional group to the glycerol skeleton, a certain functionality of the triglyceride can be imparted which is effective in delivering the desired fatty acids for maintaining healthy nutrition or treating specific diseases [3]. Lipid modification strategies for the production of functional fats and oils include chemically or lipase-catalyzed interesterification, acidolysis reactions, and genetic engineering of oilseed crops. Interesterification is used to produce fats with desirable functional and physical properties for food applications [4]. Since a physical blend of medium-chain triacylglycerols and long-chain triacylglycerols was used, with the medium-chain triacylglycerols being readily metabolized for quick energy [5], structured lipids were designed to provide simultaneous delivery of beneficial long-chain fatty acids at a slower rate and medium-chain fatty acids at a quicker rate [6]. Further, SL synthesis yields novel triacylglycerol (TAG) molecules and its derivates, such as human milk fat substitutes (HMFS), which is used in infant formula to mimic the human milk fat. Although fat accounts for only 3–5% of human milk (TAGs > 98%), it provides more than half of the energy for the growth and development of infants [7]. In addition, adding 1,3-dioleoyl-2-palmitoylglycerol (OPO) into infant formula could improve the calcium deficiency, constipation, and even bowel infarction [8]. Moreover, many phenolic lipids have also been produced, such as cocoa butter equivalents, low calorie oil, acute energy supply, and structured phospholipids. Thus, numerous structured lipids were generated from glycerol which provided important environmental benefits to the new platform products.

Phenolic acids are natural antioxidants accounting for approximately one third of the phenolic compounds in our daily diet which are widely distributed in some agricultural products, beverages, and Chinese medicinal herbs [9]. Phenolic acids and its derivatives are used in several applications in food, pharmaceutical, and cosmetic industries due to their antioxidant effects [10]. Many synthetic phenolic acids have been used as antioxidants to control lipid oxidation in lipid-based foods, and synthetic phenolics such as butylated hydroxyanisole (BHA), butylated hydroxytoluene (BHT), and tertiary butyl hydroquinone (TBHQ) are the commercially available antioxidants for many years. However, they have potential carcinogenic effects and toxicity. This has increased the utilization of natural phenolic acids to reduce the oxidation and render the health benefits [11, 12]. Nevertheless, the hydrophilic character of phenolic acids limits their effectiveness in stabilizing fats and oils as well as the applicability in food, cosmetics, and other fields. As a result of the hydrophilic nature, natural phenolic acids have been lipophilized to obtain modified amphiphilic molecules to increase their potential as antioxidants in oil-based products [13, 14]. Several reports employing chemical and enzymatic methods have been published where a number of phenolics were incorporated into different oils to lipophilize the phenolics [12, 15, 16]. With the increased interest in the phenolic lipid research, a novel phenolic lipid produced from glycerol is being pursued with improved antioxidant and biological activities.

Glycerol is a nontoxic, edible, and biodegradable compound and has over 2000 different applications [17], especially in pharmaceuticals, personal care, foods, and cosmetics. With a focus on recent developments in the conversion of glycerol into value-added chemicals, phenolic compounds conjugated with structured lipids would be interesting to study, which can expand the application of the phenolic acids and make full use of glycerol. Effective methods have been applied to produce modified phenolic acid compounds presently by lipase-catalyzed transesterification reactions in batch reactors. However, high enzyme amount, high reaction temperature, long reaction times, and low conversions often occurred. Furthermore, high vacuum was usually indispensable to remove the by-product including ethanol or water during the whole reaction course [18]. Recently, the concept of "miniaturizing biocatalysis" (i.e., microfluidic biocatalysis) was proposed and recognized as one of the priority development directions in the field of chemical engineering. Due to its excellent mass transfer characteristic, the reaction process has been accelerated, which has attracted extensive attention. Therefore, new reaction systems as well as novel lipases with better catalytic properties should be explored.

This chapter covers a broad range of information concerning the production and applications of phenolic structured lipids produced from glycerol, including natural phenolic acids, production strategies, food and medical applications, and future prospects for research and development in this field.

## 2. Phenolic acid and its derivatives

In the traditional Chinese medicine industry, phenolic compounds have been used as antimicrobials, thickeners, and flavoring agents [19] or to maintain the color of red meats. In addition, they constitute potent preservatives as of their antioxidant activity [20]. In this regard, the structured lipids integrated with phenolics can help enhance their biological properties. Thus, the source and biological activities of phenolic acids will be covered in this sector.

### 2.1 Chemical structure

Natural phenolic acids are widely found in many medicinal plants, such as honeysuckle, pallet root, dandelion, angelica, and breviscapine of the honeysuckle family. **Figure 1** shows the structures of the reported phenolic lipids. **Table 1** shows the various phenolic structured lipids with mono-and bis-phenacyl groups.

Phenolic acid compounds are mainly composed of derivatives with benzoic acid ($C_6$-$C_1$), phenylacetic acid ($C_6$-$C_2$), and cinnamic acid ($C_6$-$C_3$) as the parent nucleus and can be divided into two categories: simple phenolic acid compounds and polyphenols, including vanillic acid, *p*-hydroxybenzoic acid, salicylic acid, gallic acid, cinnamic acid, *p*-coumaric acid, ferulic acid, 3,4-dihydroxyphenylacetic acid, and protocatechuic acid [21]. Phenolic acids are resistant to oxidation mainly due to the conjugated action of the phenolic core structure and the side chain, which can easily form a stable conjugated structure [22]. Low-molecular-weight phenolic acid antioxidants mainly include (1) $C_6$-$C_1$-type of protocatechuic acid and ferulic acid, (2) $C_6$-$C_2$ 3-hydroxyphenylacetic acid, and (3) $C_6$-$C_3$ tanshinsu, caffeic acid, ferulic acid, *p*-coumaric acid, and erucic acid. Phenolic acid derivatives are also potential resources of natural medicines, which served as lead compounds for drug synthesis. Because the most important active site is o-hydroxyl group on the benzene ring, the structural modification of phenolic acids mainly focuses on the reaction of carboxyl groups, including the synthesis of amides, amine salts, and various ester compounds [23].

### 2.2 Prepare methods

Phenolic acids are abundant in the biomass feedstock that can be derived from the processing of lignin or other by-products from agro-industrial waste. Phenolic acid can be used directly in various applications, and their value can be significantly increased when they are further modified to high value-added compounds. Enzymatic reactions including esterification and decarboxylation are important for conversion of phenolic acids, which are stable and clean without toxic waste compared to the chemical methods. The products are useful for the pharmaceutical, cosmetic, food, fragrance, and polymer industries.

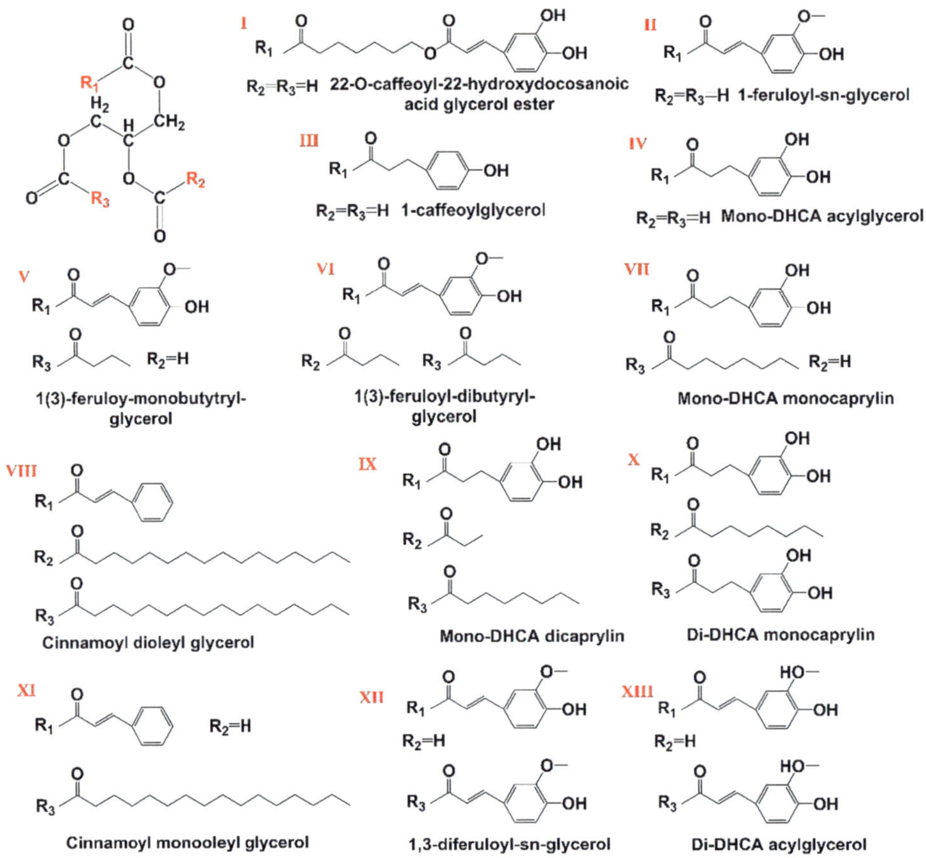


**Figure 1.**
*The structures of the reported phenolic lipids.*

### *2.2.1 Extraction and separation*

Phenolic acids are compounds that can be derived from plant biomass materials. They are the major components of lignin in lignocellulose. Depolymerization of kraft lignin using chemical, physicochemical, and biological processes can liberate substantial amounts of phenolic lipids [21, 24]. In addition to kraft lignin, depolymerization of organosolv lignin (lignin derived from environmentally friendly organic solvent-pretreated plant biomass) and extraction of lignin by aqueous formic acid can also generate many types of high-value phenolic acids [25]. By-products, residues, and wastes from fruit and vegetable industries such as from the fruit-based wine industry have significant amounts of bioactive phenolic acids which exhibit potent antioxidant activities [26]. Recently, it has been shown that derivatives of *p*-coumaric acid can be detected and isolated from palm oil waste and further processed to generate products of higher value [27].

### *2.2.2 Chemical and enzymatic synthesis*

Many phenolic acid derivatives can be potentially used as bioactive ingredients in food, cosmetic, pharmaceutical, and perfumery industries, which are from plants such as benzoic acid, salicylic acid, gallic acid, cinnamic acid, *p*-coumaric acid, caffeic acid, and ferulic acid. However, the high polarity of these compounds limits their cellular uptake, which poses a challenge for their applications. The cellular

| Type | Products | Phenolic hydroxyl position | Structural formula (Figure 1) | MW | Source | Source material | Ref. |
|---|---|---|---|---|---|---|---|
| Monophenol acyl | 1(3)-Feruloyl-dibutyryl-glycerol | *sn*-1 | VI | 407 | Enzymatic synthesis | Ethyl ferulate, tributyrin | [28] |
| | 1(3)-Feruloyl-monobutytryl-glycerol | *sn*-1 | V | 338 | Enzymatic synthesis | Ethyl ferulate, tributyrin | [28] |
| | 1-Feruloyl-*sn*-glycerol | *sn*-1 | II | 265 | Isolation | *Solanum tuberosum* (potato) | [29] |
| | Cinnamoyl dioleyl glycerol | *sn*-1 | VIII | 698 | Enzymatic synthesis | Ethyl cinnamate, triolein | [30] |
| | 22-O-Caffeoyl-22-hydroxydocosanoic acid glycerol ester | *sn*-1 | I | 368 | Isolation | Yellow cotton fiber pigment | [31] |
| | 1-Caffeoylglycerol | *sn*-1 | III | 228 | Enzymatic synthesis | Methyl caffeate, glycerol | [32] |
| | Chlorogenate fatty esters | *sn*-1 | / | $14n + 264$ | Enzymatic synthesis | 5-Caffeoylquinic acid, methyl chlorogenate | [33] |
| | Mono-DHCA dicaprylin | *sn*-1 | IX | 434 | Two-step enzymatic synthesis | Octanol, dihydrocaffeic acid, triacylglycerols | [16] |
| | Mono-DHCA monocaprylin | *sn*-1 | VII | 378 | | | |
| | Mono-DHCA acylglycerol | *sn*-1 | IV | 252 | | | |
| Bis-phenacyl | 1,3-Diferuloyl-*sn*-glycerol | *sn*-1, 3 | XII | 444 | Isolation | *Aegilops ovata* (wheat) | [34] |
| | Di-DHCA monocaprylin | *sn*-1, 3 | X | 543 | Two-step enzymatic synthesis | Octanol, dihydrocaffeic acid, triacylglycerols | [16] |
| | Di-DHCA acylglycerol | *sn*-1, 3 | XIII | 417 | | | |

*DHCA, dihydrocaffeic acid*
*$n$ = 3, 7, 11 or 15.*

**Table 1.**
*Various phenolic structured lipids with mono- and bis-phenacyl groups.*

permeability of these compounds can be increased by the esterification in order to enhance their lipophilicity. Enzymatic synthesis is the priority for esterifying phenolic lipids compared to the chemical methods using base or acid catalysts with the disadvantages of high production costs. For instance, efficient esterification of BA with heptanol to form heptyl benzoate was conducted with immobilized Novozym 435 [35]. Esterification of salicylic acid with acetic acid can be catalyzed by lipase to synthesize aspirin [36]. Decarboxylation of plant-derived phenolic lipids such as hydroxycinnamic acids is a means for generating vinylic phenol, which can be modified for use in the synthesis of food-grade flavors [37]. Thus, the efficient enzymatic method provides an efficient and sustainable way to prepare specialty chemicals and precursors for the pharmaceutical, food, fragrance, and polymer industries.

## 2.3 Biological activities

Phenolic acids in food industry have received considerable attention as powerful antioxidants to protect against the oxidative deterioration of such food components as polyunsaturated fatty acids (PUFAs) [38]. It has been proved that synthetic antioxidants have certain safety risks; the research on safe and efficient natural antioxidants has become a hot topic [39]. Additionally, phenolic acid and its derivatives have antiviral, anticancer, antioxidant, anti-inflammatory, antiaging, and other biological activities. This sector will cover the oxidation resistance, anticancer activity, and ultraviolet (UV) damage repair performance.

### *2.3.1 Oxidation resistance*

At present, the most widely natural antioxidants studied are ferulic acid, cinnamic acid, coffee-acyl quinic acid, *p*-coumaric acid, and caffeoylquinic acid, mainly due to synthetic antioxidants such as butylated hydroxyanisole and butylated hydroxytoluene which may have potential carcinogenicity [40]. With the development of synthesis of novel structural lipids, phenolic compounds have been modified to the triglyceride backbone by chemical or enzymatic methods to improve its solubility in hydrophobic media. The phenolic acid derivative has a dihydroxyphenyl structure and an ortho-phenolic hydroxyl group (1–3) in its molecular structure. The former is a common free radical scavenging structure, and the latter is easily oxidized.

### *2.3.2 Anticancer activity*

It is found that cinnamic acid derivatives have certain inhibitory effect on the proliferation of cancer cells and have certain application value in the field of anticancer. Cinnamic acid can effectively inhibit the proliferation of A-59 human lung adenocarcinoma cells [41]. Its derivatives play a significant role in inducing apoptosis of human hepatoma cells [42].

### *2.3.3 Ultraviolet damage repair performance*

In recent years, with the development of industrial production, air pollution is becoming more and more serious, leading to serious damage to the ozone layer, which makes the intensity of ultraviolet radiation gradually increased, threating the formation of skin diseases. Chemical sunscreens have problems such as poor light stability and oxidative deterioration, which can lead to skin allergies. Therefore, a new safe and efficient UV sunscreen with durable UV damage repair time and mild effect should be developed. It has been reported that ferulic acid protects against oxidative damage and apoptosis of human keratinocytes (HaCaT cells) induced by UVB, and its mechanism may be involved in the enhancement of the antioxidant activity and reduction of oxygen free radicals [43].

## 3. Enzymatic production

The enzymatic reaction is mild and highly selective, and the process route is simplified, which is the priority for the production of phenolic structured lipids. Phenolic acid can readily be esterified with glycerol skeletons to form different structured lipids via enzymatic reaction, which greatly enhance their antioxidant and functional properties. The specific selectivity of lipase makes it possible to control the position of structural lipid fatty acids. Further, solvent-free bioprocess is the priority for the efficient and environment-friendly synthesis of phenolic structured lipids [44]. This sector will cover the role of lipase, the enzymatic reaction, and the separation and characterization of phenolic structured lipids.

### 3.1 Lipase

Lipase (EC 3.1.1.3) is a general term for a class of enzymes that catalyze the hydrolysis of glycerides, which is a group of important multifunctional enzymes in the field of lipid biotechnology [45]. The most commonly used lipases in the production of phenolic structured lipids possess position or region specificity; they specifically hydrolyze the ester bonds of the triglyceride *sn*-1 and *sn*-3 and have no effect on the *sn*-2 ester bond due to steric hindrance effects. The specificity of lipase may be due to the source of lipase, the characteristic structure of the substrate, physicochemical factors at the surface, and differences in binding sites of the enzyme [46]. The transesterification of *sn*-1,3-specific lipase-catalyzed oil and triglyceride with high content of *sn*-2 unsaturated fatty acids can reduce oil saturation and increase the level of unsaturated fatty acids [47]. These kinds of lipase are mostly found in microorganisms, such as *Aspergillus niger*, *Rhizopus delemar*, *Mucor miehei*, and *Humicola lanuginosa* [48]. Lipases can be used to catalyze the resolution of optical isomers and the synthesis of chiral compounds due to their stereospecificity, which is a research hotspot in the field of enzyme engineering [49]. In addition to their high specificity and selectivity, they can remain stable under relatively high temperatures and conditions with organic solvents and no coenzymes involved, which would minimize the formation of side products and thus facilitate the subsequent separation and processing of the products [44] (**Table 2**). Thus, the use of lipases for specific lipid production is a technique that is used for many years with promising results, which allows the industry to meet the changing dietary requirement of consumers [44].

### 3.2 Reaction type

Glycerol serves as the feedstock for the production of phenolic lipids mediated by lipases through transesterification, acidolysis, alcoholysis, and interesterification [49]. Enzymatic esterification is an efficient, green and clean method for esterifying phenolic acid. The esterification of a phenolic acid compound is usually carried out by acylating a carboxyl or a hydroxyl group other than a phenolic hydroxyl one to retain its strong antioxidant capacity and enable the newly formed derivative to have good properties of fat solubility. The method for the production of structured triacylglycerols includes lipase-catalyzed esterification of fatty alcohols and phenolic acids and the transesterification of phenolic acids with acylglycerol models [44]. **Figure 2** shows the transesterification of phenolic acids with acylglycerol. It has been

| Products | Method | Substrate | Catalysts | Reactor | Reaction medium | Reaction time (h) | Substrate molar ratio | Reaction temperature (°C) | Agitation speed (rpm) | Water activity ($a_w$) | Catalyst reuse ability | Refs |
|---|---|---|---|---|---|---|---|---|---|---|---|---|
| 1(3)-Feruloyl-dibutyryl-glycerol, 1(3)-feruloyl-monobutytryl-glycerol | Transesterification | Ethyl ferulate, tributyrin | Novozym 435 | Magnetic stirrer | Toluene | 120 | 1:3 | 50 | 210 | 0.23 | 14 | [28] |
| Oleyl cinnamate | Esterification | Cinnamic acid, oleyl alcohol | Novozym 435 | Orbital shaker | Isopropanol/2-butanol | 288 | 1:6 | 55 | 150 | 0.05 | / | [50] |
| Chlorogenate fatty esters | Esterification Transesterification | 5-Caffeoylquinic acid, methyl chlorogenate | *Candida antarctica* lipase B | Orbital shaker | Solvent-free | 9 | 1:1 | 55 | 250 | 0.05 | / | [33] |
| Structured phenolic lipids (cinnamic, ferulic, sinapic and dihydrocaffeic acid) | Transesterification | Flaxseed oil cinnamic, dihydrocaffeic, 3,4-dihydroxyphenylacetic, 3,4-dimethoxybenzoic, ferulic and sinapic acids | Novozym 435 | Orbital incubator shaker | Solvent-free | 126 240 | / | 55 | 150 | / | / | [44] |
| 1-Caffeoylglycerol | Transesterification | Alkyl caffeates, glycerol | Novozym 435 | Shaken batch | Solvent-free | 10 | / | 75 | 180 | / | / | [51] |
| 1-Feruloyl-*sn*-glycerol | Esterification | Glycerol + ethyl ferulate | Novozym 435 | / | 2-Methyl-2-butanol | 168 | 1:1 | 55 | / | / | / | [10] |
| 1,3-Diferuloyl-sn-glycerol | Esterification | 4-Hydroxy-3-methoxy cinnamic acid (ethyl ferulate), soybean oil | Novozym 435 | Packed-bed column | / | 50 g/h | 1:5 | 60 | / | / | / | [43] |
| 1,3-Diferuloyl-sn-glycerol | Esterification | 4-Hydroxy-3-methoxy cinnamic acid (ethyl ferulate), soybean oil | Novozym 435 | Shaken batch | Solvent-free | 144 | 5:9 | 60 | 125 | / | / | [43] |

| Products | Method | Substrate | Catalysts | Reactor | Reaction medium | Reaction time (h) | Substrate molar ratio | Reaction temperature (°C) | Agitation speed (rpm) | Water activity ($a_w$) | Catalyst reuse ability | Refs |
|---|---|---|---|---|---|---|---|---|---|---|---|---|
| Cinnamoyl monooleyl glycerol, cinnamoyl dioleyl glycerol | Transesterification | Ethyl cinnamate, triolein | *Proteus vulgaris* K80 lipase (immobilized) | Incubator | *n*-Hexane/ toluene (85:15, v/v) | 72 | 1:6 | 35 | 210 | / | / | [30] |
| Caffeoyl monoacylglycerols, caffeoyl diacylglycerols | Transesterification | Ethyl caffeate, castor oil | Novozym 435 | Water baths with magnetic stirrers under 10 mmHg vacuum pressure | Solvent-free | 46.5 | 1:3 | 90 | / | / | / | [52] |
| Glyceryl monocaffeate | Esterification | Caffeic acid, glycerol | [$BSO_3HMIM$] TS | Oil bath | / | 2 | 1:10 | 90 | 250 | / | / | [53] |
| 1-Caffeoylglycerol | Transesterification | Methyl caffeate, glycerol | Novozym 435 | Microreactor | Chloride-urea | / | / | 65 | / | / | 20 | [32] |

**Table 2.**
*Enzymatic reactions for the production of phenolic structured lipids.*

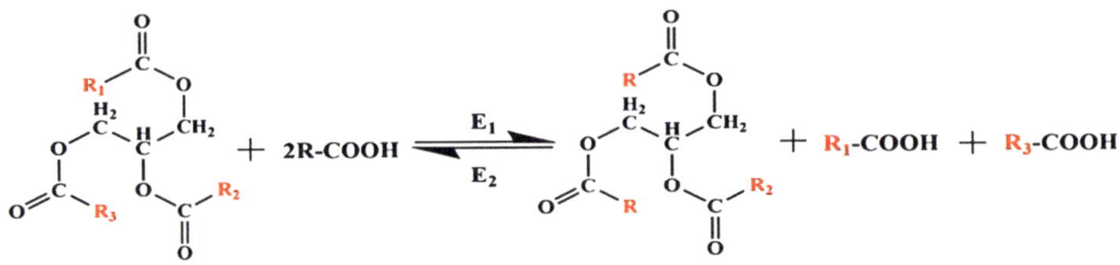


**Figure 2.**
*The transesterification of phenolic acids with acylglycerol ($R_1$, $R_2$, $R_3$, and R stand for the different fatty acids, respectively. $E_1$ and $E_2$ present the condition of the transesterification including the enzyme, temperature, and other reactions).*

reported that ethyl ferulate can be transesterified with soybean oil to synthesize mono-ferulic acid triglyceride and di-ferulic acid triglyceride [54]. Kunduru et al. performed the chemo-enzymatic synthesis of four structured triacylglycerol bearing ferulic acids as a phenolic acid at *sn*-1,3 position and found the antioxidant potency of the phenolic structured lipids measured by the Rancimat method improved compared to ferulic acid [14]. Eliza et al. incorporated ascorbic into canola acylglycerols by enzymatic transesterification, significantly improving storage and frying performance compared to the control canola oil, which offered nutraceutical ingredients for food formulation [55]. Because of the extensive activities of phenolic acid, their incorporation into triacylglycerols could potentially result in novel structured phenolic lipids having the benefits of both functional and antioxidative properties [28].

### 3.3 Reaction conditions

The organic solvent system was commonly used for the enzymatic production of phenolic lipids. Nevertheless, the use of some organic solvents may limit the acceptability of nutraceuticals and food ingredients as well as the low volumetric productivity [56]. One of the most promising novel approaches consists of using solvent-free system (SFS), which may allow the use of a smaller reaction volume and higher substrate concentrations and avoid the process of solvent recovery [57]. Feruloylated structured lipids were produced by enzymatic transesterification in solvent-free system, which achieved relatively high conversion of ethyl ferulate, reaching 98.3 ± 1.1% [15, 58]. To further improve the biological activity of phenolic lipids, ionic liquids (ILs) was also used in the enzymatic transesterification of ethyl ferulate with castor oil [59]. Castor oil-based caffeoyl structured lipids was successfully prepared which combined beneficial properties of both castor oil and caffeic acid [52]. Some other novel phenolic lipids were also produced with potential nutritional and functional benefits. Selected phenolic structured lipid synthesis was catalyzed by lipase via transesterification of 3,4-dihydroxyphenylacetic acid (DHPA) with flaxseed oil in solvent-free system. The reaction led to a significant increase in the relative proportion of linolenic acid ($C_{18}$:3 $\omega$-3) that is good for human health [60]. An efficient and solvent-free bioprocess for the synthesis of a phenolic ester of docosahexaenoic acid (DHA) was developed to expand its stability as the functional food ingredient [61].

Some other reaction variables such as reaction temperature and substrate ratio also play a crucial role in the phenolic lipid production. Temperature mainly affects the activity of lipase and the mass transfer rate in the enzymatic esterification. For instance, in the synthesis of caffeoyl structured lipids by enzymatic transesterification using monooleate as caffeoyl acceptors, with the increase of reaction temperature from 50 to 70°C, ethyl caffeate conversion reached 97.5 ± 1.9% at 70°C. Moreover, high temperatures resulted in the decrease of the reaction system viscosity, which favored the enzymatic synthesis [62]. However, temperatures above 100°C usually cause the enzyme deactivation. Similar effects of higher reaction temperature on enzyme activity were also found in other reports [15, 52]. Substrate ratio also has an impact on the enzyme activity of lipase. The high molar ratio of castor oil to ethyl ferulate from 1:1 to 1:5 led to the concomitant decrease of ethyl ferulate conversion. The reason was probably that excessive ethyl ferulate inhibits the enzyme by acidifying microaqueous phase surrounding the lipase [15, 63]. The production of such structured triacylglycerols, possessing various enrichment levels of selected fatty acids, has become an area of great interest because of their potential nutritional and functional benefits [64]; the enhancement of the solubility and miscibility properties of these novel biomolecules could increase their usefulness compared to their corresponding hydrophilic phenolic acids.

### 3.4 Separation and characterization

The separation and purification of phenolic lipids are commonly conducted by HPLC and thin-layer chromatography (TLC), further identified by FT-IR, GC-MS, atmospheric pressure chemical ionization-mass spectrometry (APCI-MS), and NMR quantitative analysis. In the lipase-catalyzed acidolysis of flaxseed oil with selected phenolic acids, the reaction components were monitored by HPLC. Cinnamic, 3,4-dihydroxyphenylacetic, and *p*-coumaric acids were monitored at 235 and 280 nm, which showed a UV-spectral scanning profile different from other phenolic acid components [65]. The transesterification reaction mixture of fish liver oil with dihydrocaffeic acid (DHCA) was analyzed qualitatively by TLC on silica gel 60 plates. The bands corresponding to the phenolic mono- and diacylglycerols visible under UV (265 nm) were recovered and separated. APCI-MS in the positive-ion mode was used to characterize the molecular structure of phenolic lipids for further analyses of the eluting peaks by HPLC [13]. The selectivity of *Candida antarctica* lipase B (CALB) was verified by elucidating the structure of the purified ester by NMR, indicating that the polyunsaturated chain was grafted on the vanillyl alcohol primary hydroxyl group, whereas the phenolic hydroxyl group remained unaffected [61].

## 4. Microreactor

Structured phenolic lipids were usually achieved through enzymatic derivatization by esterifying the carboxylic acid group with long-chain alcohols or glycerol, which could obtain an amphiphilic molecule without losing its original functional properties. Traditional methods for transesterification are simple and convenient to operate and widely used for industrial production. However, violent shaking of the mixture in the batch reactor may lead to the crack and collapse of lipase which could obviously reduce activity of lipase [66]. In addition, it is time-consuming which leads to oxidative deterioration of oil and limits the commercialization of products. Considering the violent collapse of enormous bubbles simultaneously, tremendous generation of heat and pressure occurs, which could be helpful to remove the by-product like ethanol without vacuum [67]. Thus, microreactor technology has been proposed to be beneficial for the effective production of phenolic structured lipids.

Microreactors are recognized as powerful tools for chemical synthesis. The specific surface area of microreactor is much larger than that of conventional reactor, which possesses strong heat exchange capacity and fast mass transfer rate. Moreover, the reaction time was greatly reduced, and the products and substrates can be easily separated. Therefore, microreactors are suitable for reactions with a severe reaction process or a high-temperature requirement.

### 4.1 Reactor type and characteristics

Propyl caffeate was achieved in the microreactor with 1-heptyl-methylimidazolium bis(trifluoromethylsulfonyl)imide [$C_7$mim] [$Tf_2N$] as a cosolvent. The yield of 99.50% was achieved, while the yield in the conventional reactor was 98.50%, and the reaction time was 9/10 shorter than that of the conventional reactor (24 h) [68]. Moreover, human milk fat-style structured triacylglycerols were produced from microalgal oil in a continuous microfluidic reactor packed with immobilized lipase, which obtained high conversion efficiency with reaction time being reduced by eight times [63]. The packed bed reactor was built in a stainless steel plate with lipozyme RM IM used as a biocatalyst, and *n*-hexane was employed as the medium (**Figure 3**). Further, the packed bed reactor was designed with polydimethylsiloxane which has a good chemical inertness with dimension 10 × 0.9 × 75 mm (W × H × L), respectively. One inlet and one outlet were set in both ends of the microreactor to ensure the substrate was fed from one inlet and flow out of the reactor from an outlet which was on the other side of the reactor. The bottom of the groove was packed with Novozym 435; the homogeneous solutions of methyl caffeate, glycerin, and cosolvent were pumped into the reactor for the synthesis of 1-caffeoylglycerol (1-CG) [32]. Due to the advantages of strong heat exchange capacity and fast mass transfer rate in microreactors, the high yield of the 1-caffeoylglycerol was achieved with high reaction rate, low energy consumption, and short time and continuous production. Thus, microreactors represent a convenient and cost-saving method to produce phenolic lipids using microfluidic biocatalysis.

### 4.2 Reaction conditions

The reaction in microreactors is commonly affected by flow rate, temperature, and substrate concentration. With lower flow rate, the substrate will be fully mixed and collided with the enzyme which makes the substrate contact with the active site of the lipase sufficient, and a high yield of 1-caffeoylglycerol (1-CG) was achieved [32]. The viscosity at a low temperature is very high, due to the high boiling point of glycerin. It

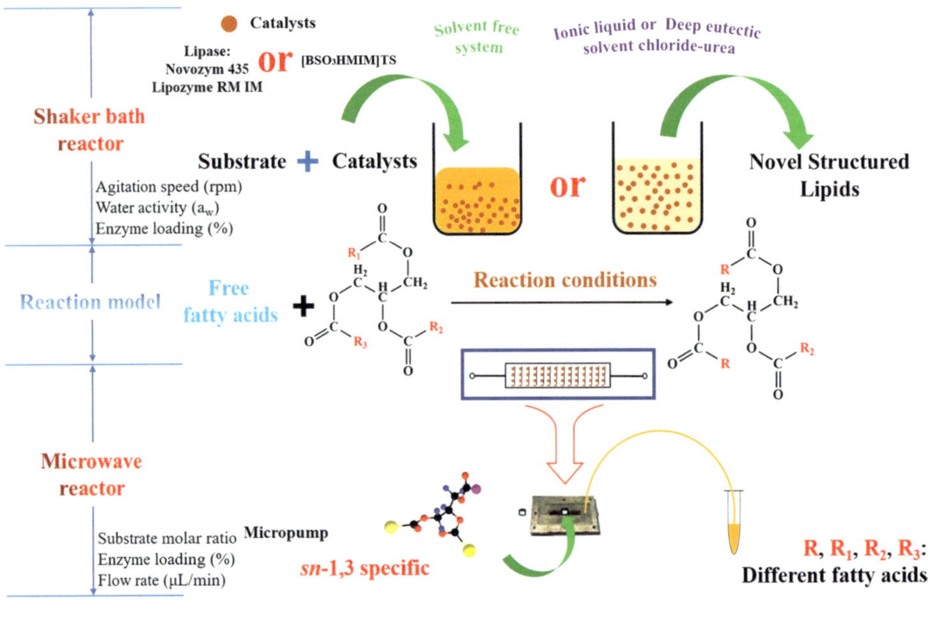


**Figure 3.**
*Synthesis of human milk fat-style structured lipids in a continuous microfluidic reactor.*

is important to reduce the viscosity of the system and increase the mass transfer rate by raising the temperature in microreactors. Substrate concentration is expected to effect the incorporation of glycerol ester with fatty acid. Even though higher substrate concentrations can promote more incorporation of polyunsaturated fatty acids (PUFAs) to triacylglycerols at the initial reaction, the high level molar ratios of substrate may inhibit lipase activity and also complicate the downstream purification [69].

### 4.3 Kinetic analysis

Enzyme kinetics is an important means to evaluate different reactor performances. Kinetic modeling plays a role as an engineering practice in accelerating enzymatic reactions which indicates the behavior of substrate and enzyme [70]. In addition, enzyme kinetic modeling can explore pathways and reaction mechanisms of complex macromolecular substrates using many parameters prior to developing innovative process to ensure stability and desired efficiency [71]. Herein, the enzyme kinetic modeling was used to evaluate catalytic efficiency of enzyme and mass transfer in microreactors. In our previous study, the value of kinetic parameters ($K_{m(app)}$) which is relative to the flow rate was calculated for the aim of kinetic study in continuous-flow microreactors [72]. The effect of flow rate on enzyme kinetics is usually investigated using the Lilly-Hornby model [73]. In the production of human milk fat-style structured triglycerides, at the lower flow rate, a lower $K_{m(app)}$ value was obtained, which shows that the acyl donors and enzyme were in sufficient contact with one another [63]. In the enzymatic synthesis of 1-caffeoylglycerol (1-CG), ping-pong bi-bi model was used for the kinetic analysis. To further reveal the effect of internal mass on the reaction, the modeling of fluid flow in microchannels can be achieved by using a formulation in which a common flow field is shared by all of the phases. Numerical simulation liquid flow was applied in the microreactor with the COMSOL Reaction Engineering Lab 3.5a [32]. In summary, kinetic approach can analyze reaction parameters and explain the high efficiency of microreactor.

Thus, the microfluidic technology is promising for modified functional lipid production. However, there are still few reports about microfluidic bioconversion technology employed in the phenolic structured lipid production. More research on phenolic structured lipid production involving microfluidic technology stays explored to provide a cost-effective approach for producing high-value coproducts.

## 5. Applications

Glycerol is emerging as a versatile bio-feedstock for the production of a variety of chemicals, polymers, and fuels. New catalytic conversions of glycerol have been applied for the synthesis of products whose use ranges from everyday life to the fine-chemical industry. In addition, phenolic acids show a great potential ability of antitumor, antioxidant, antibacterial, and anti-ultraviolet activity due to the unique structure. Thus, in order to utilize potential ability of phenolic acids to meet the demand of cancer treatment and food antioxidants, artificial transformation of its structure using glycerol as feedstock is daily crucial and has been the mainstream research direction in recent years. However, there are few reports on application of the structure triglycerides containing phenolic acids; recent studies have focused on the antioxidant and anticancer aspects of mofetil. The research about repair ability of structure triglycerides on UV damage of cells has also been a hot topic. Therefore, this sector will focus on the application of phenolic acid structural lipid on the antioxidant and anticancer aspects of mofetil as well as the ultraviolet damage repair performance.

### 5.1 Antioxidants

Antioxidant capacity of phenolic acids ester is usually evaluated in the following three ways, such as 2,2-diphenyl-1-picrylhydrazyl free radical (DPPH) scavenging, antioxidant potency in lipid matrix using Rancimat, and the rate of inhibition of autoxidation of linoleic acid in micelles. Four compounds of structured phenolic lipids of varying chain lengths were synthesized to evaluate their antioxidant ability [14]. It was found that after the combination of ferulic acid and glyceride, the antioxidant capacity of ferulic acid glyceride was significantly improved. With ferulic acid as reference, the oxidation time was increased from 12.9 to 15.05 h. In addition, sinapic acid, which was considered as one of the dietary phenolic acids, also is evaluated in the study of Gaspar et al. Alkyl ester sinapates (linear alkyl esters) present almost the same antioxidant activity, albeit slightly lower, compared with the parent compound (sinapic acid) [74]. It was also found that the addition of an alkyl ester side chain shows positive effect on the utilization as an antioxidant in a more lipophilic medium via improving the partition coefficient. Furthermore, ester derivatives of ferulic acid also show the superior antioxidant to the parent ferulic acid [75]. In our previous study, we have synthesized 1-caffeoylglycerol (1-CG), and the ability of 1-CG to scavenge DPPH free radicals and repair UV damage of HaCaT cell was studied, which showed the most effective antioxidant function on DPPH and repair function on UV damage of HaCaT cells compared to methyl caffeate and caffeic acid. Thus, the combining phenolic acid with triglycerides helps phenolic acids to exert antioxidant functions in fat-soluble foods and pharmaceuticals.

### 5.2 Anticancer agents

Phenolic acids have been reported to have the ability of anticancer via inhibiting the growth of tumor cell or selecting inducers of cell death. The gray GM(0, N) approach was employed to analyze the structure activity relationship of phenolic acid phenethyl esters on oral and human breast cancers [76]. **Figure 4** shows the chemical structures of phenolic acid phenethyl esters, and among that structure $R_1$ is the most important functional group which has a great influence on the cytotoxicity to tumor cell (SAS, OEC-M1, MCF-7). Thus, the ability of inhibiting the

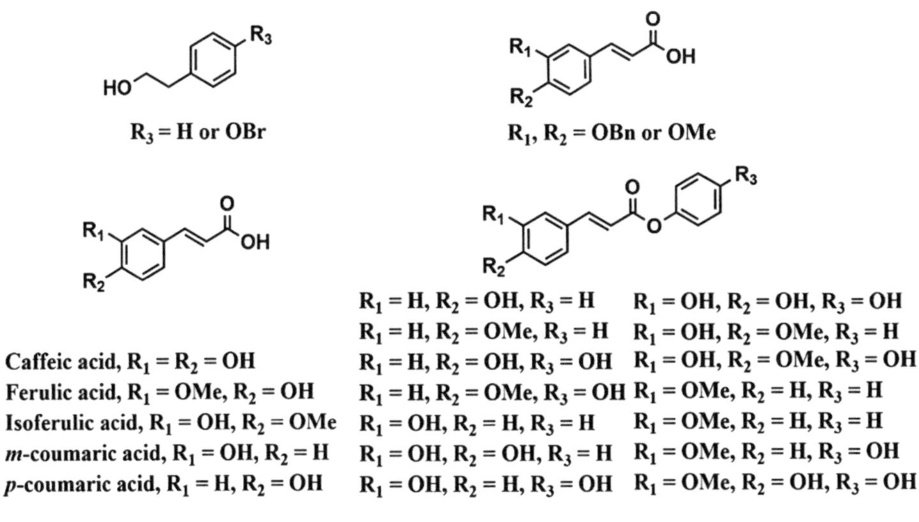


**Figure 4.**
*Chemical structures of phenolic acid phenethyl esters.*

growth of the tumor cell could be influenced by the variable of phenolic acid ester. In addition, phenolic acid ester can also induce the death of the cancer cell. It is reported that 13-D have the ability of selectively inducing apoptosis in white blood cancers [77]. These phenolic acid esters not only have selective osmotic effects but also block the cell cycle, and its target compounds are localized in the nucleus and cytoplasm. Furthermore, it is reported that the viability of Detroit 562 cells could be significantly influenced by caffeic acid phenethyl ester [77]. Thus, the phenolic acid ester presented the capability of inhibiting the growth and inducing the apoptotic response of cancer cells.

## 6. Conclusions

New applications of glycerol as a low-cost feedstock for functional structured lipids have been found, which indicated converting glycerol into commercially valued products. In order to utilize potential ability of phenolic acids, they have been added to structured lipids produced from glycerol to impart even more functionality. Enzymatic esterification reaction was mostly used to enhance the liposolubility of phenolic acids. Microfluidic technology has also been an effective tool for the phenolic lipid production. Whether it is through improvement in functionality or physical properties of a food or the medicinal properties, phenolic structured lipids definitely provide attributes that consumers will find valuable. Therefore, it is important that further research is conducted that will allow for better understanding and more control over the various esterification processes and reduction in costs associated with large-scale production of phenolic structured lipids.

## Acknowledgements

This study was financially supported by the Key Research and Development Program (Modern Agriculture) of Jiangsu Province (BE2017322), the Key Research and Development Program (Modern Agriculture) of Zhenjiang City (NY2017010), the Six Talent Peaks Project of Jiangsu Province (2015-NY-018), the 333 High-Level Talent Training Project of Jiangsu Province (Year 2018), the Shen Lan Young scholars program of Jiangsu University of Science and Technology (Year 2015), the Postgraduate Research & Practice Innovation Programs of Jiangsu Province (SJKY19_2670, KYCX18_2305).

## Conflict of interest

The authors have declared no conflicts of interest.

## Author details

Jun Wang[1,2]*, Linlin Zhu[1], Jinzheng Wang[1], Yan Hu[1] and Shulin Chen[3]

1 School of Biotechnology, Jiangsu University of Science and Technology, Zhenjiang, PR China

2 Sericultural Research Institute, Chinese Academy of Agricultural Sciences, Zhenjiang, PR China

3 Department of Biological Systems Engineering, Washington State University, Pullman, Washington, USA

*Address all correspondence to: wangjun@just.edu.cn

## References

[1] Tan HW, Aziz AA, Aroua MK. Glycerol production and its applications as a raw material: A review. Renewable and Sustainable Energy Reviews. 2013;**27**:118-127

[2] Mostafa NA, Maher A, Abdelmoez W. Production of mono-, di-, and triglycerides from waste fatty acids through esterification with glycerol. Advances in Bioscience and Biotechnology. 2013;**4**(09):900

[3] Vu P, Shin J, Lee Y, Nam H, Lee J, Akoh CC, et al. Development and characterization of structured lipids containing capric and conjugated linoleic acids as functional dietary lipid molecules. International Journal of Food Sciences and Nutrition. 2009;**59**(2): 95-104. DOI: 10.1080/09637480701461531

[4] Salazar R, Arámbula-Villa G, Hidalgo FJ, Zamora R. Structural characteristics that determine the inhibitory role of phenolic compounds on 2-amino-1-methyl-6-phenylimidazo[4,5-b]pyridine (PhIP) formation. Food Chemistry. 2014;**151**:480-486. DOI: 10.1016/j.foodchem.2013.11.105

[5] Jala RCR, Hu P, Yang T, Jiang Y, Zheng Y, Xu X. Lipases as biocatalysts for the synthesis of structured lipids. Methods in Molecular Biology. 2012;**861**:403. DOI: 10.1007/978-1-61779-600-5_23

[6] Willis WM, Lencki RW, Marangoni AG. Lipid modification strategies in the production of nutritionally functional fats and oils. CRC Critical Reviews in Food Technology. 1998;**38**(8):639-674. DOI: 10.1080/10408699891274336

[7] Wei W, Feng Y, Zhang X, Cao X, Feng F. Synthesis of structured lipid 1,3-dioleoyl-2-palmitoylglycerol in both solvent and solvent-free system. LWT-Food Science and Technology. 2015;**60**(2):1187-1194

[8] Innis SM, Dyer RP, Diersenschade D. Palmitic acid is absorbed as sn-2 monopalmitin from milk and formula with rearranged triacylglycerols and results in increased plasma triglyceride sn-2 and cholesteryl ester palmitate in piglets. Journal of Nutrition. 1995;**125**(1):73-81

[9] Balasundram N, Sundram K, Samman S. Phenolic compounds in plants and Agri-industrial by-products: Antioxidant activity, occurrence, and potential uses. Food Chemistry. 2006;**99**(1):191-203. DOI: 10.1016/j.foodchem.2005.07.042

[10] Compton DL, Laszlo JA, Evans KO. Antioxidant properties of feruloyl glycerol derivatives. Industrial Crops and Products. 2012;**36**(1):217. DOI: 10.1016/j.indcrop.2011.09.009

[11] Jiankang W, Fereidoon S. Acidolysis of p-coumaric acid with omega-3 oils and antioxidant activity of phenolipid products in in vitro and biological model systems. Journal of Agricultural and Food Chemistry. 2013;**62**(2): 454-461. DOI: 10.1021/jf404140v

[12] Wang J, Shahidi F. Antioxidant activity of monooleyl and dioleyl p-coumarates in in vitro and biological model systems. European Journal of Lipid Science and Technology. 2014;**116**(4):370-379. DOI: 10.1002/ejlt.201300348

[13] Sabally K, Karboune S, St-Louis R, Kermasha S. Lipase-catalyzed synthesis of phenolic lipids from fish liver oil and dihydrocaffeic acid. Biocatalysis and Biotransformation. 2009;**25**(2-4):211-218. DOI: 10.1080/10242420701379916

[14] Reddy KK, Shanker KS, Ravinder T, Prasad RBN, Kanjilal S. Chemo-enzymatic synthesis and evaluation of novel structured phenolic lipids as potential lipophilic antioxidants.

European Journal of Lipid Science and Technology. 2010;**112**(5):600-608. DOI: 10.1002/ejlt.200900200

[15] Sun S, Zhu S, Bi Y. Solvent-free enzymatic synthesis of feruloylated structured lipids by the transesterification of ethyl ferulate with castor oil. Food Chemistry. 2014;**158**:292-295. DOI: 10.1016/j.foodchem.2014.02.146

[16] Yang Z, Feddern V, Glasius M, Zheng G, Xu X. Improved enzymatic production of phenolated acylglycerols through alkyl phenolate intermediates. Biotechnology Letters. 2011;**33**(4):673-679. DOI: 10.1007/s10529-010-0486-3

[17] Centi G, van Santen RA. Catalysis for Renewables: From Feedstock to Energy Production. Weinheim, Germany: Wiley-VCH; 2008

[18] Xu C, Zhang H, Shi J, Zheng M, Xiang X, Huang F, et al. Ultrasound irradiation promoted enzymatic alcoholysis for synthesis of monoglyceryl phenolic acids in a solvent-free system. Ultrasonics Sonochemistry. 2018;**41**:120-126. DOI: 10.1016/j.ultsonch.2017.09.016

[19] Ayala-Zavala JF, Vega-Vega V, Rosas-Domínguez C, Palafox-Carlos H, Villa-Rodriguez JA, Siddiqui MW, et al. Agro-industrial potential of exotic fruit byproducts as a source of food additives. Food Research International. 2011;**44**(7):1866-1874. DOI: 10.1016/j.foodres.2011.02.021

[20] Senanayake SPJN. Green tea extract: Chemistry, antioxidant properties and food applications—A review. Journal of Functional Foods. 2013;**5**(4):1529-1541. DOI: 10.1016/j.jff.2013.08.011

[21] Tinikul R, Chenprakhon P, Maenpuen S, Chaiyen P. Biotransformation of plant-derived phenolic acids. Biotechnology Journal. 2018;**13**(6):1700632. DOI: 10.1002/biot.201700632

[22] Pourova J, Kottova M, Voprsalova M, Pour M. Reactive oxygen and nitrogen species in normal physiological processes. Acta Physiologica. 2010;**198**(1):15-35. DOI: 10.1111/j.1748-1716.2009.02039.x

[23] Heleno SA, Martins A, Queiroz MJRP, Ferreira ICFR. Bioactivity of phenolic acids: Metabolites versus parent compounds: A review. Food Chemistry. 2015;**173**:501-513. DOI: 10.1016/j.foodchem.2014.10.057

[24] Shi Y, Chai L, Tang C, Yang Z, Zhang H, Chen R, et al. Characterization and genomic analysis of Kraft lignin biodegradation by the beta-proteobacterium Cupriavidus basilensis B-8. Biotechnology for Biofuels. 2013;**6**(1):1

[25] Duan J, Huo X, Du WJ, Liang JD, Wang DQ, Yang SC. Biodegradation of Kraft lignin by a newly isolated anaerobic bacterial strain, Acetoanaerobium sp. WJDL-Y2. Letters in Applied Microbiology. 2016;**62**(1):55-62

[26] Akyol H, Riciputi Y, Capanoglu E, Caboni MF, Verardo V. Phenolic compounds in the potato and its byproducts: An overview. International Journal of Molecular Sciences. 2016;**17**(6):835

[27] Pinthong C, Phoopraintra P, Chantiwas R, Pongtharangkul T, Chenprakhon P, Chaiyen P. Green and sustainable biocatalytic production of 3,4,5-trihydroxycinnamic acid from palm oil mill effluent. Process Biochemistry. 2017;**63**:122-129

[28] Zheng Y, Wu XM, Branford-White C, Ning X, Quan J, Zhu LM. Enzymatic synthesis and characterization of novel feruloylated lipids in selected organic media. Journal of Molecular Catalysis B Enzymatic. 2009;**58**(1):65-71. DOI: 10.1016/j.molcatb.2008.11.005

[29] Graça J, Pereira H. Suberin structure in potato periderm: Glycerol, long-chain monomers, and glyceryl and feruloyl dimers. Journal of Agricultural and Food Chemistry. 2000;**48**(11):5476-5483. DOI: 10.1021/jf0006123

[30] Jo JC, Kim HK. Production of cinnamoyl lipids using immobilized proteus vulgaris K80 lipase and an evaluation of their antioxidant activity. Journal of Molecular Catalysis B: Enzymatic. 2016;**129**:54-60. DOI: https://doi.org/10.1016/j.molcatb.2016.04.008

[31] Ma M, Hussain M, Memon H, Zhou W. Structure of pigment compositions and radical scavenging activity of naturally green-colored cotton fiber. Cellulose. 2016;**23**(1):955-963. DOI: 10.1007/s10570-015-0830-9

[32] Liu X, Meng XY, Xu Y, Dong T, Zhang DY, Guan HX, Zhuang Y, Wang J. Enzymatic synthesis of 1-caffeoylglycerol with deep eutectic solvent under continuous microflow conditions. Biochemical Engineering Journal. 2019;**142**:41-49. DOI: 10.1016/j.bej.2018.11.007

[33] López Giraldo LJ, Laguerre M, Lecomte J, Figueroa-Espinoza M, Barouh N, Baréa B, Villeneuve P. Lipase-catalyzed synthesis of chlorogenate fatty esters in solvent-free medium. Enzyme and Microbial Technology. 2007;**41**(6):721-726. DOI: https://doi.org/10.1016/j.enzmictec.2007.06.004

[34] Cooper R, Gottlieb HE, Lavie D. New phenolic diglycerides from *Aegilops ovata*. Phytochemistry. 1978;**17**(9):1673-1675. DOI: https://doi.org/10.1016/S0031-9422(00)94673-9

[35] Giunta D, Sechi B, Solinas M. Novozym-435 as efficient catalyst for the synthesis of benzoic and (hetero) aromatic carboxylic acid esters. Tetrahedron. 2015;**71**(18):2692-2697. DOI: 10.1016/j.tet.2015.03.036

[36] Jamwal S, Dharela R, Gupta R, Ahn J, Chauhan GS. Synthesis of crosslinked lipase aggregates and their use in the synthesis of aspirin. Chemical Engineering Research and Design. 2015;**97**:159-164. DOI: 10.1016/j.cherd.2014.09.010

[37] Mishra S, Sachan A, Vidyarthi AS, Sachan SG. Transformation of ferulic acid to 4-vinyl guaiacol as a major metabolite: A microbial approach. Reviews in Environmental Science and Bio/Technology. 2014;**13**(4):377-385. DOI: 10.1007/s11157-014-9348-0

[38] Masuda T, Akiyama J, Takeda Y, Maekawa T, Sone Y. Identification of the antioxidation reaction products from a sinapic ester in a lipid oxidation system. Bioscience, Biotechnology, and Biochemistry. 2014;**73**(3):736-739. DOI: 10.1271/bbb.80636

[39] Ekiert H, Piekoszewska A, Muszyńska B, Baczyńska S. Accumulation of p-coumaric acid and other bioactive phenolic acids in vitro culture of ruta grvaeolens ssp. divaricata (tenore) gams. Medicina Internacia Revuo. 2014;**102**(26):24-31

[40] Magnani C, Isaac VLB, Correa MA, Salgado HRN. Caffeic acid: A review of its potential use in medications and cosmetics. Analytical Methods. 2014;**6**(10):3203-3210. DOI: 10.1039/C3AY41807C

[41] Song HY, Liu YK, Feng JT, Cui JF, Dai Z, Zhang LJ, Feng JX, Shen HL, Tang ZY. Proteomic analysis on metastasis-associated proteins of human hepatocellular carcinoma tissues. Chinese Journal of Hepatology. 2005;**13**(5):331. DOI: 10.1007/s00432-005-0044-x

[42] Putt KS, Nesterenko V, Dothager RS, Hergenrother PJ. The compound 13-D selectively induces apoptosis in white blood cancers versus other cancer cell types. Chembiochem.

2010;**7**(12):1916-1922. DOI: 10.1002/cbic.200600228

[43] Compton DL, Laszlo JA. 1,3-Diferuloyl-sn-glycerol from the biocatalytic transesterification of ethyl 4-hydroxy-3-methoxy cinnamic acid (ethyl ferulate) and soybean oil. Biotechnology Letters. 2009;**31**(6):889. DOI: 10.1007/s10529-009-9952-1

[44] Sorour N, Karboune S, Saint-Louis R, Kermasha S. Lipase-catalyzed synthesis of structured phenolic lipids in solvent-free system using flaxseed oil and selected phenolic acids as substrates. Journal of Biotechnology. 2012;**158**(3):128-136. DOI: 10.1016/j.jbiotec.2011.12.002

[45] Kuoa TC, Shawb JF, Leea GC. Improvement in the secretory expression of recombinant candida rugosa lipase in pichia pastoris. Process Biochemistry. 2015;**50**(12):2137-2143. DOI: 10.1016/j.procbio.2015.09.013

[46] Chen H, Meng X, Xu XQ, Liu WB, Li SY. The molecular basis for lipase stereoselectivity. Applied Microbiology & Biotechnology. 2018;**102**(1-4):1-9. DOI: 10.1007/s00253-018-8858-z

[47] Lin TJ, Chen SW, Chang AC. Enrichment of n-3 PUFA contents on triglycerides of fish oil by lipase-catalyzed trans-esterification under supercritical conditions. Biochemical Engineering Journal. 2006;**29**(1):27-34. DOI: 10.1016/j.bej.2005.02.035

[48] Heubi JE, Schaeffer D, Ahrens RC, Sollo N, Strausbaugh S, Graff G, et al. Safety and efficacy of a novel microbial lipase in patients with exocrine ancreatic insufficiency due to cystic fibrosis: A randomized controlled clinical trial. Journal of Pediatrics. 2016;**176**:156-161. DOI: 10.1016/j.jpeds.2016.05.049

[49] Li D, Wang W, Liu P, Xu L, Faiza M, Yang B, et al. Immobilization of *Candida antarctica* lipase B onto ECR1030 resin and its application in the synthesis of n-3 PUFA-rich triacylglycerols. European Journal of Lipid Science & Technology. 2017;**119**(12):1700266. DOI: 10.1002/ejlt.201700266

[50] Lue B, Karboune S, Yeboah FK, Kermasha S. Lipase-catalyzed esterification of cinnamic acid and oleyl alcohol in organic solvent media. Journal of Chemical Technology & Biotechnology. 2005;**80**(4):462-468. DOI: 10.1002/jctb.1237

[51] Meng XY, Xu Y, Wu JX, Zhu CT, Zhang DY, Wu GH, et al. Enzymatic synthesis and antioxidant activity of 1-Caffeoylglycerol prepared from alkyl caffeates and glycerol. Journal of the American Oil Chemists' Society. 2018;**95**(2):149-159. DOI: 10.1002/aocs.12036

[52] Sun SD, Wang P, Zhu S. Enzymatic incorporation of caffeoyl into castor oil to prepare the novel castor oil-based caffeoyl structured lipids. Journal of Biotechnology. 2017;**249**:66-72. DOI:10.1016/j.jbiotec.2017.03.022

[53] Sun SD, Hou XB, Hu BX. Synthesis of glyceryl monocaffeate using ionic liquids as catalysts. Journal of Molecular Liquids. 2017;**248**:643-650. DOI: https://doi.org/10.1016/j.molliq.2017.10.102

[54] Wang J, Gu S, Pang N, Wang F, Wu FA. A study of esterification of caffeic acid with methanol using p-toluenesulfonic acid as a catalyst. Journal of the Serbian Chemical Society. 2013;**78**(7):1023-1034. DOI: 10.2298/JSC120802101W

[55] Gruczynska E, Przybylski R, Aladedunye F. Performance of structured lipids incorporating selected phenolic and ascorbic acids. Food Chemistry. 2015;**173**:778-783. DOI: 10.1016/j.foodchem.2014.10.122

[56] Martínez I, Markovits A, Chamy R, Markovits A. Lipase-catalyzed solvent-free transesterification of wood sterols. Applied Biochemistry & Biotechnology.

2004;**112**(1):55-62. DOI: 10.1385/abab:112:1:55

[57] Lima RN, Silva VR, Santos LDS, Bezerra DP, Soares MBP, Porto ALM. Fast synthesis of amides from ethyl salicylate under microwave radiation in a solvent-free system. Rsc Advances. 2017;7(89):56566-56574. DOI: 10.1039/C7RA11434F

[58] Sun S, Song F, Bi Y, Yang G, Liu W. Solvent-free enzymatic transesterification of ethyl ferulate and monostearin: Optimized by response surface methodology. Journal of Biotechnology. 2013;**164**(2):340-345. DOI: 10.1016/j.jbiotec.2013.01.013

[59] Sun SD, Zhu S. Enzymatic preparation of castor oil-based feruloylated lipids using ionic liquids as reaction medium and kinetic model. Industrial Crops & Products. 2015;**73**:127-133. DOI: 10.1016/j.indcrop.2015.04.019

[60] Sorour N, Karboune S, Saint-Louis R, Kermasha S. Enzymatic synthesis of phenolic lipids in solvent-free medium using flaxseed oil and 3,4-dihydroxyphenyl acetic acid. Process Biochemistry. 2012;**47**(12):1813-1819. DOI: 10.1016/j.procbio.2012.06.020

[61] Roby MH, Allouche A, Dahdou L, Castro VCD, Silva PHAD, Targino BN, Huguet M, Paris C, Chrétien F, Guéant RM. Enzymatic production of bioactive docosahexaenoic acid phenolic ester. Food Chemistry. 2015;**171**:397-404. DOI: 10.1016/j.foodchem.2014.09.028

[62] Sun SD, Hu BX. Enzymatic preparation of novel caffeoyl structured lipids using monoacylglycerols as caffeoyl acceptor and transesterification mechanism. Biochemical Engineering Journal. 2017;**124**:78-87. DOI: 10.1016/j.bej.2017.05.002

[63] Wang J, Liu X, Wang XD, Dong T, Zhao XY, Zhu D, Mei YY, Wu GH. Selective synthesis of human milk fat-style structured triglycerides from microalgal oil in a microfluidic reactor packed with immobilized lipase. Bioresource Technology. 2016;**220**:132-141. DOI: 10.1016/j.biortech.2016.08.023

[64] Halldorsson A, Magnusson CD, Haraldsson GG. Chemoenzymatic synthesis of structured triacylglycerols by highly regioselective acylation. Tetrahedron. 2003;**59**(46):9101-9109. DOI: 10.1016/j.tet.2003.09.059

[65] Safari M, Safari M, Karboune S, St-Louis R, Kermasha S. Enzymatic synthesis of structured phenolic lipids by incorporation of selected phenolic acids into triolein. Biocatalysis. 2009;**24**(4):272-279. DOI: 10.1080/10242420600658410

[66] Dimakou CP, Kiokias SN, Tsaprouni IV, Oreopoulou V. Effect of processing and storage parameters on the oxidative deterioration of oil-in-water emulsions. Food Biophysics. 2007;**2**(1):38-45. DOI: 10.1007/s11483-007-9027-6

[67] Bansode SR, Rathod VK. Ultrasound assisted lipase catalysed synthesis of isoamyl butyrate. Process Biochemistry. 2014;**49**(8):1297-1303. DOI: 10.1016/j.procbio.2014.04.018

[68] Hua Z, Baker GA, Shaletha H. New eutectic ionic liquids for lipase activation and enzymatic preparation of biodiesel. Organic & Biomolecular Chemistry. 2011;**9**(6):1908-1916. DOI: 10.1039/c0ob01011a

[69] Zhong NJ, Gui ZY, Xu L, Huang JR, Hu K, Gao YQ, Zhang X, Xu ZB, Su JY, Li B. Solvent-free enzymatic synthesis of 1, 3-diacylglycerols by direct esterification of glycerol with saturated fatty acids. Lipids in Health and Disease. 2013;**12**(1):65. DOI: 10.1186/1476-511X-12-65

[70] Vasic-Racki D, Bongs J, Schörken U, Sprenger GA, Liese A. Modeling of

reaction kinetics for reactor selection in the case of L-erythrulose synthesis. Bioprocess and Biosystems Engineering. 2003;**25**(5):285-290

[71] Ülger C, Takaç S. Kinetics of lipase-catalysed methyl gallate production in the presence of deep eutectic solvent. Biocatalysis and Biotransformation. 2017;**35**(6):407-416. DOI: 10.1080/10242422.2017.1359573

[72] Wang J, Gu SS, Cui HS, Wu XY, Wu FA. A novel continuous flow biosynthesis of caffeic acid phenethyl ester from alkyl caffeate and phenethanol in a packed bed microreactor. Bioresource Technology. 2014;**158**:39-47. DOI: 10.1016/j.biortech.2014.01.145

[73] Lilly MD, Hornby WE, Crook EM. The kinetics of carboxymethylcellulose-ficin in packed beds. Biochemical Journal. 1966;**100**(3):718

[74] Gaspar A, Martins M, Silva P, Garrido EM, Garrido J, Firuzi O, Miri R, Saso L, Borges F. Dietary phenolic acids and derivatives. Evaluation of the antioxidant activity of sinapic acid and its alkyl esters. Journal of Agricultural and Food Chemistry. 2010;**58**(21):11273-11280. DOI: 10.1021/jf103075r

[75] Borgohain R, Handique JG, Guha AK, Pratihar S. A theoretical study on antioxidant activity of ferulic acid and its ester derivatives. Journal of Theoretical and Computational Chemistry. 2016;**15**(04):1650028. DOI: 10.1142/s0219633616500280

[76] Lee Y. Structure activity relationship analysis of phenolic acid phenethyl esters on oral and human breast cancers: The grey GM (0, N) approach. Computers in Biology and Medicine. 2011;**41**(7):506-511. DOI: 10.1016/j.compbiomed.2011.04.016

[77] Dziedzic A, Kubina R, Kabała-Dzik A, Tanasiewicz M. Induction of cell cycle arrest and apoptotic response of head and neck squamous carcinoma cells (Detroit 562) by caffeic acid and caffeic acid phenethyl ester derivative. Evidence-based complementary and alternative medicine. 2017;**2017**:1-10. DOI: 10.1155/2017/6793456

Chapter 5

# Glycerol Transformation to Value-Added 1,3-Propanediol Production: A Paradigm for a Sustainable Biorefinery Process

*Shanthi Priya Samudrala*

## Abstract

The impact of diminishing fossil fuel resources, rising environmental issues as well as the global demand for energy, fuels and chemicals has significantly directed to the use of renewable biomass for sustainable production of fuels and chemicals. Glycerol, a three carbon feedstock, is one of the most promising biomass resources which at present is obtained as a by-product in large quantities during the biodiesel production. This stimulated a lot of interest in developing new valorization technologies to produce high-value tonnage chemicals from glycerol by sustainable processes such as oxidation, dehydration, hydrogenolysis, steam reforming, carboxylation, acetalization, esterification and chlorination. In this chapter, we intend to focus on the hydrogenolysis of glycerol which produces important commodity chemicals such as propanediols, propanols and ethylene glycol. In particular, the selective hydrogenolysis of glycerol to 1,3-propanediol performed in both liquid phase and vapor phase reaction processes is described. Furthermore, the most significant progress in the development of the catalytic materials for glycerol hydrogenolysis including the reaction pathways is herein summarized.

**Keywords:** glycerol, hydrogenolysis, biomass, biodiesel, 1,3-propanediol, metal-acid catalyst

## 1. Introduction

The consideration of renewable energy resources and sustainable processes for energy, fuel and chemical production is the most important research targeted areas of this century. Currently around 90% of all chemicals are synthesized from fossil fuels or petrochemicals. However, to combat the issues of climate change, depletion of oil resources as well as its escalating prices, the investigation of the potential renewable sources to meet the future energy requirements has been a perpetual study driving researcher's attention. Consequently, there is a pressing need to develop alternative processes for the sustainable ways of producing fuels and chemicals. In this regard, utilizing "biomass as feedstock" is well thought out as an outstanding sustainable renewable energy source for fuel and chemical production, as a substitute to fossil fuel energy.

### 1.1 Biomass

In the recent past, biomass has been recognized as the most widespread renewable energy source and a potential substitute to fossil fuels that could be regenerated faster. It is extensively used as a building block for fuels, fuel additives and fine chemicals. Biomass refers to plant- and animal-derived biological material. By using different processes such as thermal, chemical, biological and electrochemical methods, biomass resources can be converted to valuable fuels and chemicals [1]. Biomass can be chemically converted to other usable forms of energy like methane gas or transportation fuels like ethanol and biodiesel. Vegetable oils and animal fats are used to produce 'biodiesel', whereas the fermentation of corn and sugarcane produces 'ethanol'. Unlike fossil fuels, carbon dioxide released into the atmosphere by biomass combustion is absorbed by the plants as the regrowth of the biomass crops takes shorter time and hence results in a closed carbon cycle with net zero carbon emission. Therefore, biomass is regarded as the naturally occurring clean and efficient renewable energy carbon resource used as a substitute for fossil fuels.

### 1.2 Biodiesel

Biodiesel is a renewable, potentially viable environmentally friendly diesel fuel, which directly replaces conventional petroleum diesel. Given its non-toxic and biodegradable nature, biodiesel offers several benefits including better combustion emission profile, with low greenhouse gas emissions. The enhanced lubricating properties that can extend the engine life, high flash point (150°C) and material compatibility made biodiesel a benign transportation fuel and a better substitute to petroleum-based fuel. It can either be blended with petroleum in different proportions or can be used directly in its pure form in the diesel engines. Base-catalyzed transesterification process of canola oil/soya bean oil, animal fats or cooking oil (triglycerides) with alcohol produces biodiesel (fatty acid methyl esters) and glycerol (valuable by-product) (**Figure 1**).

Biodiesel production has shown a steady growth from 2000 to 2010 rising from less than 1 million tons to 10 million tons, respectively. For every 1 tonne of biodiesel synthesized, 100 kg of glycerol is cogenerated, approximately 10 wt% of total product which is fairly greater than that produced from soap manufacturing process. Due to the rapid growth in biodiesel production, a glut of glycerol has been created, and subsequently the market value of glycerol is dropped abruptly. One of the key problems challenged by biodiesel companies is the disposal of glycerol, which is quite expensive. Now the major question arises! How the glycerol

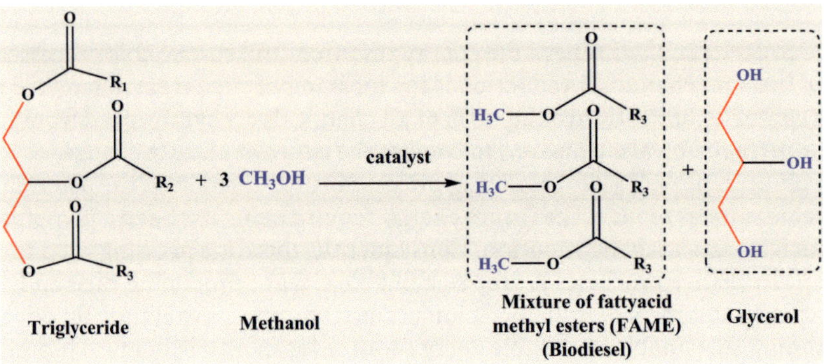


**Figure 1.**
*Biodiesel production through transesterification of triglyceride.*

is separated and valorized from biodiesel production? Therefore, to contribute to the cost affordability of biodiesel processes and for sustainable development, it is imperious to develop methods of producing value-added chemicals from glycerol. In this context, the excess glycerol derived from biodiesel process could be utilized as a renewable feedstock for substituting fossil-derived chemicals, and research is being directed worldwide to utilize glycerol as a chemical building block [2].

### 1.3 Bio-glycerol

Glycerol or glycerine also known as propane-1, 2, 3-triol is a simple trihydroxy sugar alcohol primarily used as a sweetening agent, solvent, pharmaceutical agent and emollient. It is a colorless, odorless, viscous liquid miscible in water. Glycerol is non-toxic and hygroscopic in nature. Purified glycerin is used in the production of various foods, beverages, pharmaceuticals, cosmetics and other personal care products. Glycerol is produced by saponification process (traditional soap manufacture) and obtained as a major by-product in biodiesel process (fatty acid ester production) (**Figure 2**). The convenient method for the production of glycerol is biological fermentation and hydrogenolysis of glucose. Synthetic glycerol is prepared from propylene, but this method is at present economically unappealing because of rising crude oil prices [4].

It is projected that by 2020, glycerol production will rise six times more than request. The crude glycerol from biodiesel process is of variable quality comprising a mixture of methanol, unreacted mono-, di- and triglycerides, fatty acids, water, inorganic salts, methyl esters and a range of organic constituents with an estimated 50% purity and has a low market value. The purification processes of glycerol are expensive, and it is being disposed as a waste by small biodiesel plants. This remains problematic because the methanol content of glycerol is considered as harmful waste. Currently, energy production by incineration is one method to remove surplus glycerol. In terms of both environmental protection and economic benefit, the value addition of glycerol is of great importance to further enhance the cost efficacy of biodiesel processes.

### 1.4 Valorization of bio-glycerol

With glycerol offered as a low cost, large-volume bio-feedstock, consumption of surplus glycerol is a prerequisite for the commercial sustainability of biodiesel processes. Glycerol is one among the top 12 platform chemicals from biomass, compelled by its non-toxic, edible, biodegradable properties as well as multifunctional structure. Glycerol is a highly attractive molecule used to produce variety of valuable chemical intermediates. In order to provide new applications for glycerol valorization, chemical

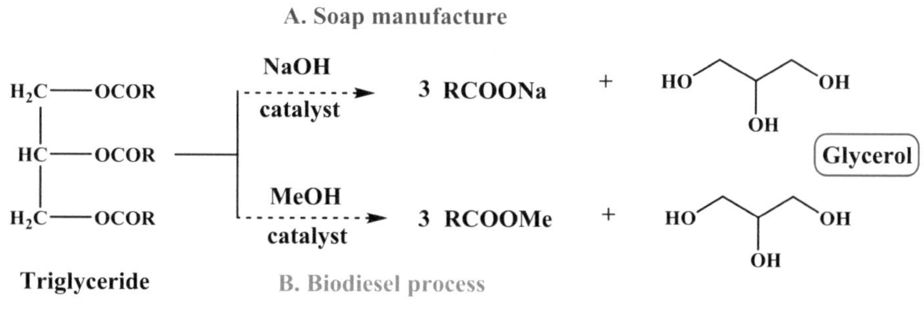


**Figure 2.**
*Glycerol production by saponification and biodiesel processes [3].*

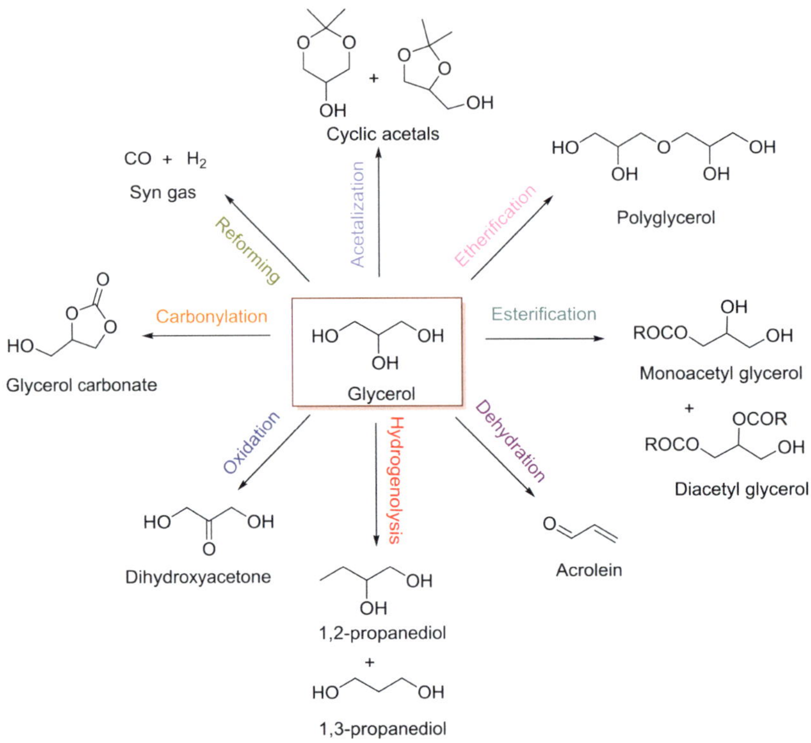


**Figure 3.**
*Different chemical transformations of glycerol to value-added chemicals.*

catalysis is an important approach to green chemical processes. In this regard, significant research on the chemical transformation of glycerol into high value-added specialty chemicals has been developed during the past decade [5].

Glycerol is a highly functionalized molecule used as a precursor for the production of wide range of commodity chemicals such as 1,2-propanediol, 1,3-propanediol, ethylene glycol, propanols, hydrocarbons, acrolein, dihydroxyacetone, glyceric acid, syngas, hydrogen, glyceryl ethers, glyceryl esters, glycerol carbonate, 1,3-dichloro propanol, polyglycerols as well as acetals and ketals of glycerol by means of several different methods such as fermentation, hydrogenolysis, pyrolysis, oxidation, etherification, dehydration, esterification, carboxylation, halogenation, polymerization and glycerol acetalization [6, 7] (**Figure 3**). Furthermore, glycerol has received a great attention as a 'green solvent' in synthetic organic chemistry because of its exceptional physical and chemical properties. Its compatibility with most organic/inorganic compounds, high boiling point, negligible vapor pressure, easy dissolution and nonhazardous nature, simple handling and storage provides an innovative way to revalorize glycerol and to be used as sustainable reaction medium [8].

## 2. Hydrogenolysis of glycerol

Hydrogenolysis reaction is a class of reduction which involves dissociation of chemical bonds (C—C or C—O) in an organic compound and subsequent addition of hydrogen to the resultant molecular fragments. C—C hydrogenolysis reactions have gained commercial importance in petroleum refineries, a method to produce lower hydrocarbons. On the other hand, C—O hydrogenolysis reduces the oxygen content and is an important area of research for biomass conversion to fuels and chemicals (**Figure 4**). C—O hydrogenolysis is highly favorable industrial relevant route of glycerol conversion processes as glycerol is rich in oxygen functionality.

Glycerol hydrogenolysis is a complex reaction involving many reaction pathways giving rise to valuable products. At first, C—O hydrogenolysis of glycerol yields propanediols (1,2-propanediol and 1,3-propanediol). Subsequent C—O hydrogenolysis gives rise to propanols (1-propanol and 2-propanol) and finally propane, whereas C—C hydrogenolysis of glycerol or the C—C hydrogenolysis of the products derived from C—O hydrogenolysis gives degradation products such as ethylene glycol, ethanol, methanol, ethane and methane. The reaction routes and products from glycerol hydrogenolysis are shown in **Figure 5**. 1,2-propanediol (1,2-PDO), 1,3-propanediol (1,3-PDO), ethylene glycol (EG), 1-propanol (1-PrOH) and 2-propanol (2-PrOH) are industrially important products of glycerol hydrogenolysis. Therefore, to increase the cost-effectiveness of biodiesel industry, the catalytic hydrogenolysis of glycerol appears to be an attractive option. Furthermore, the products obtained from glycerol hydrogenolysis can certainly replace the chemical compounds produced from non-renewable sources.

## 2.1 Propanediols (1,2-propanediol and 1,3-propanediol)

Propanediols are the most important value-added chemicals obtained from glycerol hydrogenolysis. 1,2-propanediol (1,2-PDO) is a medium-value commodity

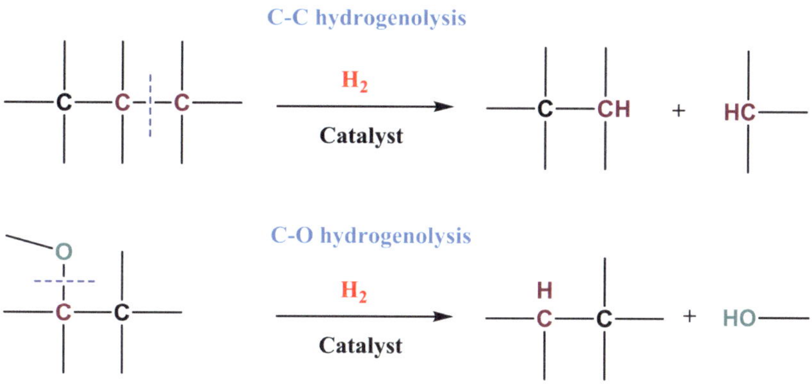


**Figure 4.**
*Reaction pathways of hydrogenolysis.*

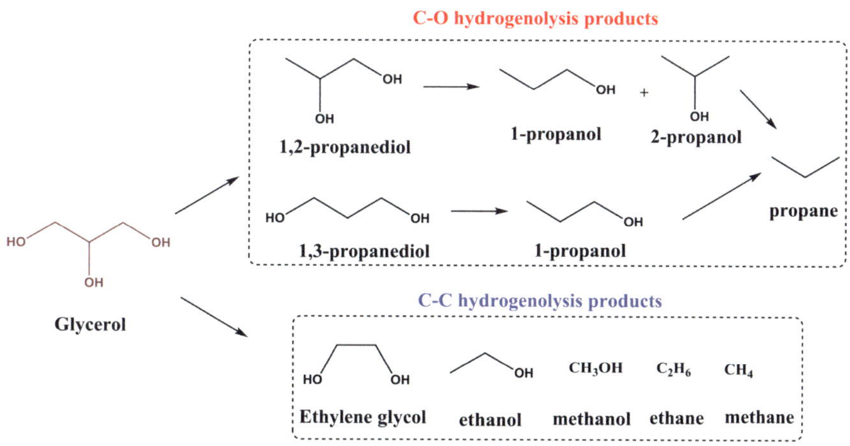


**Figure 5.**
*Products obtained from glycerol hydrogenolysis.*

chemical used as a component in paints, liquid detergents, cosmetics, food, animal feed, personal care products, tobacco humectants, flavors and fragrances. Besides being an antifreeze coolant and a de-icing agent [9], it is used as raw material for the synthesis of polyester resins, in the fiber manufacture and in pharmaceutical production. Conventionally, 1,2-propanediol is produced by hydroperoxide process or the chlorohydrin process from propylene oxide. 1,3-propanediol (1,3-PDO) is other valuable product of glycerol hydrogenolysis used as an important monomer in the manufacture of polymethylene terephthalate (PTT) or polypropylene terephthalate (PPT), a biodegradable polyester used in textile and carpet manufacturing. It finds applications in the manufacture of cosmetics, personal care, cleaning, lubricants, medicines, and in the synthesis of heterocyclic compounds in addition to engine coolants, food and beverages, de-icing fluids, water-based inks, heat transfer fluids and unsaturated polyester resins [10]. The global 1,3-PDO production is estimated to reach $621.2 million by 2021 growing at an annual growth rate of 10.4%.

## 2.2 Reaction pathways and mechanistic studies of glycerol hydrogenolysis

Glycerol is a trihydroxy compound with two primary hydroxyl groups and a secondary hydroxyl group. The hydrogenolysis reaction of glycerol involves C—O or C—C chemical bond cleavage in the presence of catalysts and hydrogen. 1,2-PDO is produced by the hydrogenolysis of primary (1°) hydroxyl groups of glycerol, whereas the hydrogenolysis of secondary (2°) hydroxyl group gives 1,3-PDO. Successive hydrogenolysis of 1,2-PDO produces monohydroxy 1-PrOH and 2-PrOH and eventually propane. The consecutive removal of the remaining —OH from 1,3-PDO could yield 1-PrOH first and then propane. The reaction mechanisms of glycerol hydrogenolysis and the products obtained mostly depend on the operating conditions of the reaction systems, nature of catalytic materials like metal properties and acidity or basicity of catalysts [11]. The generally accepted reaction mechanisms of glycerol hydrogenolysis are illustrated below.

### *2.2.1 Two-step mechanism*

The two-step mechanism of glycerol hydrogenolysis is the dehydration-hydrogenation route that involves acid-catalyzed dehydration of glycerol first and subsequent hydrogenation of intermediate in the presence of metallic sites to yield the final product (**Figure 6**). This mechanism is highly favorable under acidic conditions. Acid-catalyzed dehydration of 1° hydroxyl groups of glycerol

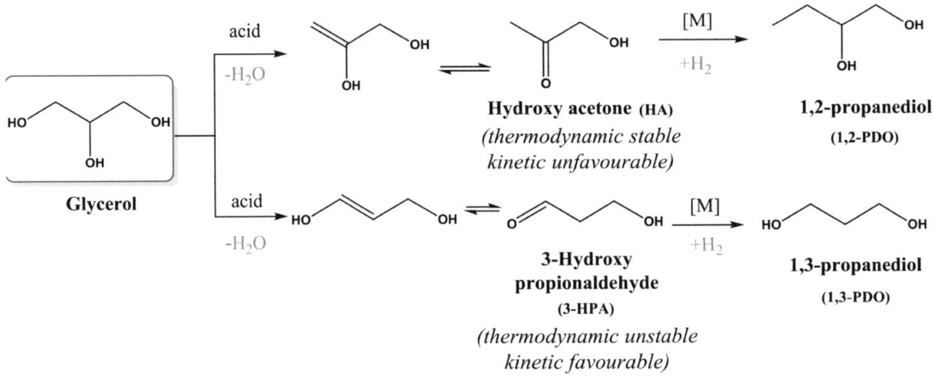


**Figure 6.**
*Dehydration-hydrogenation (two-step) mechanism of glycerol hydrogenolysis [11].*

generates hydroxyacetone (acetol) followed by hydrogenation over metal to 1,2-PDO. Similarly, 1,3-PDO is formed by the acid-catalyzed dehydration of 2° hydroxyl group of glycerol to form 3-hydroxy propionaldehyde (3-HPA) and subsequent hydrogenation over metal. However, 3-hydroxy propionaldehyde, though kinetically favorable, is thermodynamically less stable than hydroxyacetone. Furthermore, double dehydration of glycerol will produce acrolein, whereas double dehydration followed by hydrogenation gives rise to the formation of 1-PrOH and 2-PrOH [11].

#### 2.2.2 Three-step mechanism

Another route of glycerol hydrogenolysis is dehydrogenation-dehydration-hydrogenation which is a three-step mechanism, more prevalent under basic conditions [11]. It involves initial dehydrogenation of glycerol to form glyceraldehyde and 2-hydroxy acrolein by glyceraldehyde dehydration followed by hydrogenation of 2-hydroxy acrolein into 1,2-PDO (**Figure 7**). On the other hand, 1,2-PDO could also be generated by the keto-enol tautomerization of 2-hydroxy acrolein into pyruvaldehyde which hydrogenates to acetol and further hydrogenation to 1,2-PDO. Pyruvaldehyde also undergoes oxidation to yield lactic acid, and on the other hand, ethylene glycol (EG) formation can be explained through the retro-aldol reaction of glyceraldehyde to produce glycolaldehyde followed by its hydrogenation.

#### 2.2.3 Direct route

A direct glycerol hydrogenolysis reaction mechanism (**Figure 8**) was demonstrated over specific catalysts such as Ir-ReOx/$SiO_2$. In this mechanism, the glycerol adsorbed on the interface of metal species generates 2,3-dihydroxypropoxide and 1,3-dihydroxypropoxide. The hydride attack facilitates the dissociation of C—O bonds in the alkoxides, and final hydrolysis yields respective products [12].

### 2.3 Role of acidic sites in the selective hydrogenolysis of glycerol to propanediols

The properties of catalyst systems and reaction conditions effect the selective cleavage of primary or secondary hydroxyl group from glycerol. It is evident from previous research that 1,2-propanediol (1,2-PDO) and 1,3 -propanediol (1,3-PDO) are produced

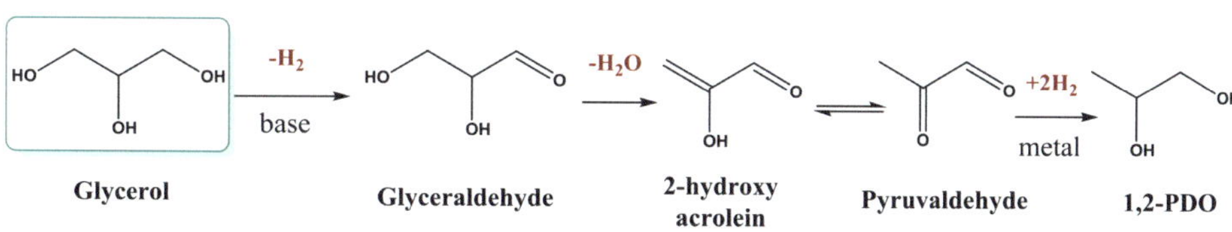


**Figure 7.**
*Dehydrogenation-dehydration-hydrogenation (three-step) mechanism of glycerol hydrogenolysis [11].*

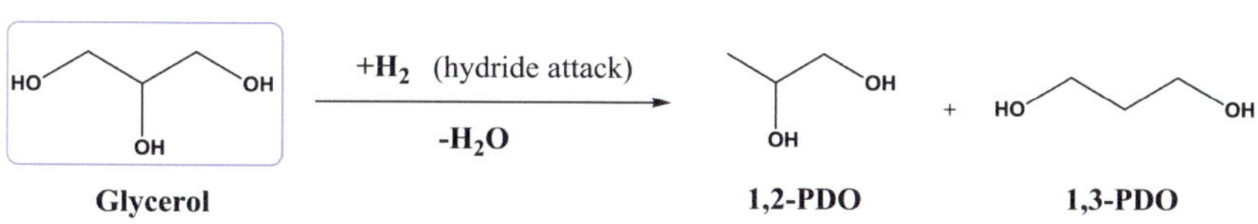


**Figure 8.**
*Direct hydrogenolysis mechanism of glycerol hydrogenolysis [3].*

from glycerol hydrogenolysis by acid-catalysed dehydration of glycerol to form acetol and 3-hydroxypropaldehyde (3-HPA) intermediates and successive hydrogenation on metal sites. Acrolein is produced by the double dehydration of glycerol over strong acid sites. Hence, surface acidity or acid strength of the catalyst is an important parameter in directing the product distribution. Indeed, the nature of acid sites also plays a vital role in determining the selective formation of propanediols from glycerol. Brønsted acid site favors the abstraction of 2° hydroxyl group of glycerol to generate 1,3-PDO via 3-HPA, while Lewis acid sites assist in the dehydration of 1° hydroxyl group of glycerol to form 1,2-PDO through the acetol intermediate (**Figure 9**) [13].

The nature of acidic sites governs the reaction mechanism of selective glycerol conversion to propanediols which is quite interesting and important in designing the appropriate catalytic materials (**Figure 10**). The primary or secondary —OH group of glycerol is attacked by the proton from a Brønsted acidic site (**Figure 10A**) without any steric hindrance. However, protonation at secondary —OH group of glycerol generates secondary carbocation which is more stable than primary carbocation resulted from protonation of primary —OH group. It should be noted that 3-HPA is thermodynamically unstable; however, its formation is kinetically more favorable over the formation of acetol Or hydroxyacetone. Hence, reaction proceeds via secondary —OH group protonation followed by dehydration to form highly unstable 3-HPA intermediate that quickly hydrogenates on metallic sites to produce 1,3-PDO. The immediate hydrogenation of 3-HPA is essential to stop further dehydration of 3-HPA to yield acrolein as final product. The Lewis acid sites of the catalysts possess empty orbital to accommodate the lone pair of electrons either from primary or secondary—OH groups of glycerol by coordinate covalent bond (CCB). This CCB seems to appear like a complex due to the direct interaction of the metal empty orbital (from the catalyst surface) with —OH groups of glycerol. However, the complex is preferentially formed with less steric hindered primary —OH group of glycerol than that of secondary —OH group. Therefore, the interaction of Lewis acid site with primary —OH group followed by dehydration results in a thermodynamically stable hydroxyacetone intermediate which is further hydrogenated to produce 1,2-PDO (**Figure 10B**). The mechanism of glycerol hydrogenolysis based on the nature of acidic sites is well in accordance with the reaction mechanism proposed by Alhanash et al. [14].

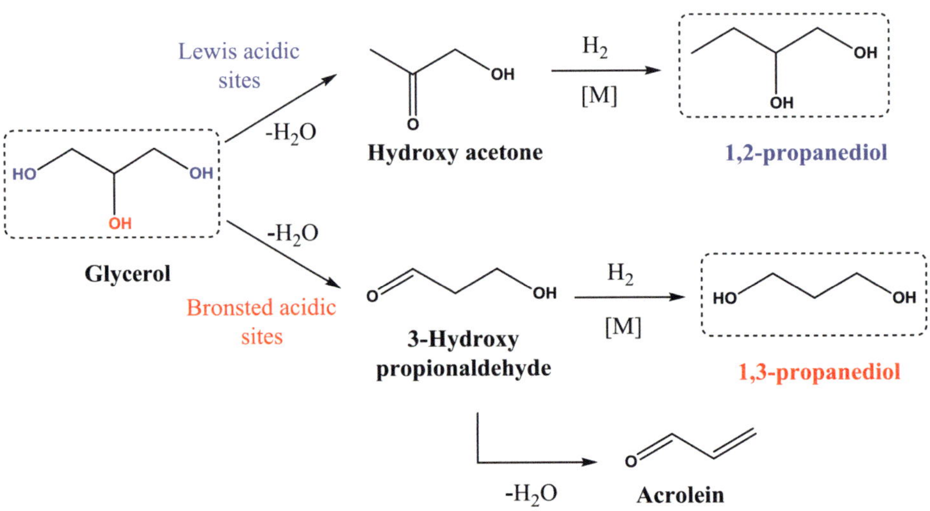


**Figure 9.**
*Selective production of propanediols from glycerol.*

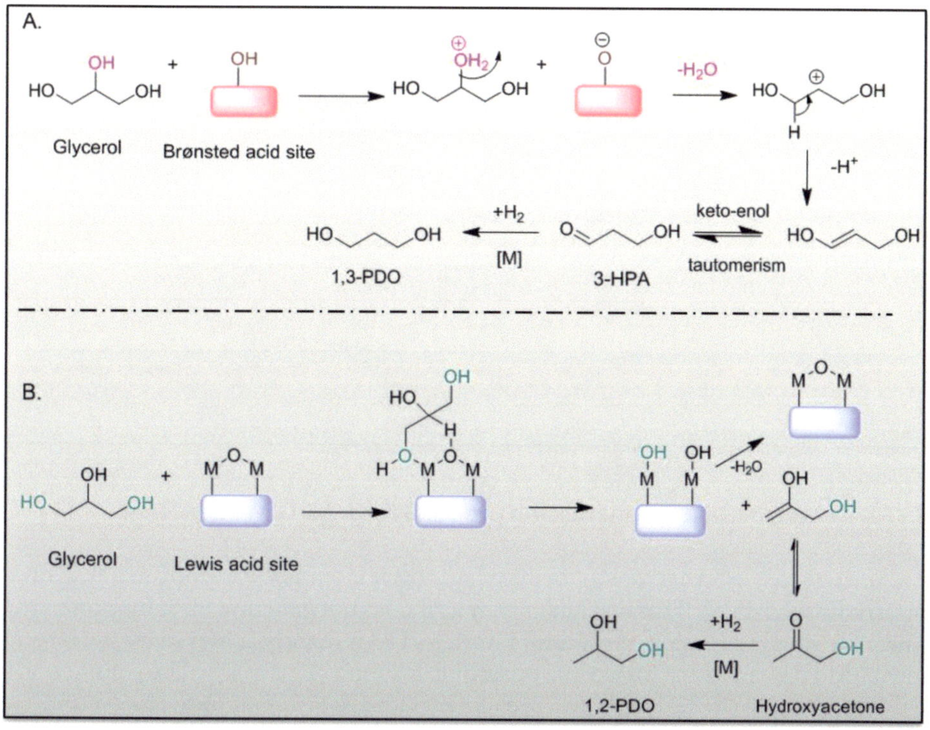


**Figure 10.**
*Selective production of propanediols directed by the nature of acidic sites [14].*

## 3. Selective conversion of glycerol to 1,3-propanediol

In glycerol hydrogenolysis, the selective cleavage of one of the C—O (primary or secondary) bonds over C—C cleavage is highly challenging and depends on the operating reaction conditions as well as the nature of catalytic material in particular acidic properties of the catalyst [11]. The type of acidic sites (Lewis acid sites or Brønsted acid sites) plays an important role in directing towards the desired product formation. 1,3-propanediol is an industrially important commodity chemical and is a component of industrial polyesters such as Dupont's Sorona®, Shell Chemical's CorterraTM and CDP Natureworks®. Three different methods including chemical, microbial and catalytic methods have so far been the commonly employed methodologies for the production of 1,3-PDO.

### 3.1 Chemical method (conventional routes) for 1,3-PDO production

There are two well-known chemical methods for the synthesis of 1,3-PDO (**Figure 11**). The first method employs 'ethylene oxide' as the feedstock and commonly known as 'Shell' route of 1,3-PDO synthesis. It is a two-step process involving hydroformylation of ethylene oxide to 3-hydroxypropionaldehyde (3-HPA) and subsequent hydrogenation of 3-HPA to 1,3-PDO. The second method of 1,3-PDO synthesis is 'Degussa-DuPont' route and uses acrolein as the feedstock [15]. The hydration of acrolein gives 3-HPA which is further hydrogenated to 1,3-PDO. However, the chemical methods of 1,3-PDO synthesis are petroleum-based methods and have many disadvantages such as high pressure, high temperature and catalysts. Consequently, the costs of 1,3-PDO production becomes very high.

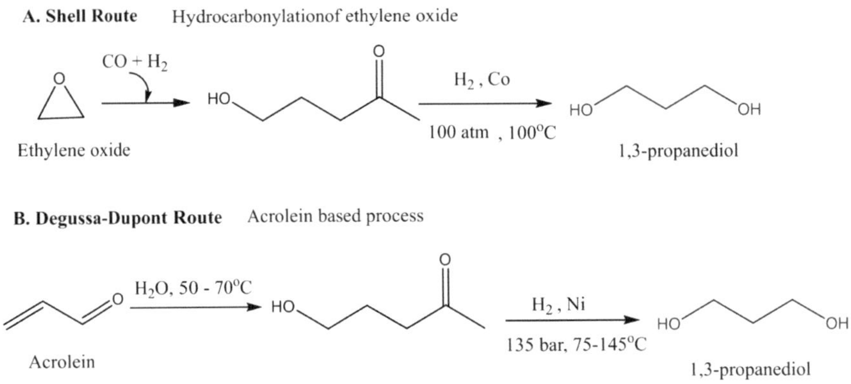


**Figure 11.**
*Traditional synthetic methods for the production of 1,3-propanediol [15].*

## 3.2 Microbial method (fermentation process) for 1,3-PDO production

An attractive alternative for chemical synthesis is a microbial conversion of raw materials to 1,3-PDO. This method is easy and does not generate toxic by-products. The biological conversion of glycerol into 1,3-PDO is mainly achieved by the fermentation process using anaerobic or aerobic bacteria [12, 16, 17]. The commonly used microbes in the fermentation process include *Clostridium*, *Enterobacter* and *Lactobacillus* [18]. The biological conversion of glycerol in a bacterial cell is carried out at functional temperature (37°C) and normal atmospheric pressure. In this process (**Figure 12**), glycerol is first dehydrated to 3-HPA in the presence of B12-dependent enzyme, and then 1,3-PDO is produced by the reduction of 3-HPA in the presence of NADH-oxido reductase. However, the energy consumption and cost of the equipment in the fermentation process are relatively high.

## 3.3 Catalytic method (homogeneous and heterogeneous catalysis) for 1,3-PDO production

In view of shortcomings of chemical and biological methods, the most economically attractive approach would be the development of new, cost-competitive process that utilizes renewable resources as feedstocks for green and sustainable chemical technologies. Catalytic method is one such process which not only represents milestones towards the goal of reducing fossil fuel dependency and greenhouse gas emissions but also has the potential to provide substantial energy savings. As glycerol hydrogenolysis to propanediols proceeds via a dehydration-hydrogenation mechanism, typically the reaction requires a bifunctional catalyst where metals (transition or noble metals) are used for hydrogenation and acidic or basic support materials can

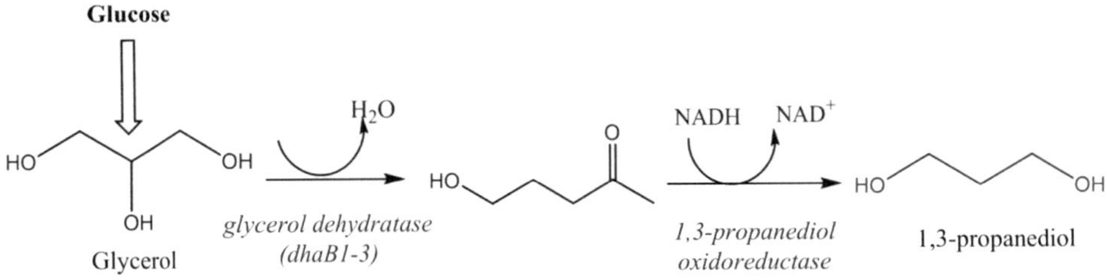


**Figure 12.**
*Microbial production of 1,3-propanediol.*

be used for dehydration. The literature reviews suggests that 1,2-PDO formation can be achieved by a simple acidic or basic catalyst; however, the formation of 1,3-PDO is challenging and definitely requires a cocatalyst that acts as a Brønsted acid additive [11]. The hydrogenolysis of glycerol has been investigated by quite a few researchers in the recent past for glycerol hydrogenolysis both in batch and continuous flow reactors using homogeneous as well as heterogeneous catalytic systems. Catalysts based on various transition and noble metals such as Cu, Ni, Co, Pd, Ru, Rh, Ir, Re, Ag, Au and Pt have been tested in glycerol hydrogenolysis [19, 20]. This chapter highlights the catalytic materials employed for the selective hydrogenolysis of glycerol to 1,3-PDO till date and summarizes all different catalyst systems used.

### *3.3.1 Homogeneous catalytic processes*

The homogeneous catalytic hydrogenolysis of glycerol has been demonstrated by Celanese and Shell corporations using Rh and Pd metal complexes in the late 1980s and 1990s [21, 22]. Later, Bullock et al. described the dehydroxylation of glycerol in sulfolane at 110°C and 52 bar pressure over a ruthenium catalyst, but 1,2-PDO and 1,3-PDO were produced in very low yields (<5%) [23]. The results of homogeneous catalytic systems have been represented in **Table 1**. Nonetheless, the major drawback of these homogeneous catalytic processes is the separation/recovery of catalyst from product and corrosion of the reactor due to reaction medium. Furthermore, glycerol hydrogenolysis using homogeneous catalysts leads to a range of by-products. Apparently, homogeneous catalytic processes end up with hazardous environmental problems, and in efforts to address these problems, researchers paid significant attention on heterogeneous catalysts.

### *3.3.2 Heterogeneous catalytic systems*

The problematic separation of homogeneous catalysts from their products and rather their instability led to the development of successful, industrially applicable heterogeneous catalysts with high levels of activity, selectivity and catalyst stability. Different classical heterogeneous metal catalysts Cu, Ru, Rh, Ir, Pt and Pd supported on a wide range of qualitatively different carriers and in presence of Brønsted acids as cocatalysts were, so far, widely employed for direct hydrogenolysis of glycerol to obtain 1,3-propanediol. Wide-ranging reaction parameters including temperature (around 180–350°C) and hydrogen pressure (1–9 MPa) are typically applied. As shown in the mechanism of glycerol hydrogenolysis via dehydration-hydrogenation and dehydrogenation-dehydration-hydrogenation

| Author (year) | Catalyst | Solvent/additive | Temp (°C) | P (MPa) | Yield of 1,3-PDO (%) | Ref. |
|---|---|---|---|---|---|---|
| Che (1985) | $Rh(CO)_2(acac)$ | MPI[a]/$H_2WO_4$ | 200 | 32 syngas | 21 | [21] |
| Shell (1998) | Pd complex[b] | Sulfolane-water/MSA[c] | 140 | 6 syngas | 30.8[d] | [22] |
| Bullock (2001) | Ru-complex[e] | Sulfolane | 110 | 5.2 $H_2$ | <5 | [23] |

[a]*MPI, 1-methyl-2-pyrrolidinone.*
[b]*Pd-BCPE (BCPE, 1,2-bis(1,5-dicyclohexylphosphino)ethane).*
[c]*Methane sulfonic acid.*
[d]*Selectivity.*
[e]$[\{Cp^*Ru(CO)_2\}_2Cu\text{–}H]^+OTf^-$ ($Cp^* = \eta^5\text{-}C_5Me_5$; $OTf = OSO_2CF_3$).

**Table 1.**
*Glycerol hydrogenolysis to 1,3-propanediol over homogeneous catalytic systems.*

routes, for selective production of 1,3-PDO from glycerol, addition of a suitable cocatalyst that can act as a Brønsted acid is highly essential. From the earlier reports, it is clearly evident that Brønsted acid catalysts have been successfully employed and found to be highly active for 1,3-PDO production. The Brønsted acidity in the catalyst could be generated either from addition of a cocatalyst/additive to the catalyst or directly from the support material chosen. Supported noble metal (Pt, Ir, Rh) catalysts with a cocatalyst or a Brønsted acid additive have been proven to be more efficient catalytic materials for the selective production of 1,3-PDO from glycerol hydrogenolysis. The catalytic materials so far investigated for the selective hydrogenolysis of glycerol to 1,3-propanediol can be categorized into four different types based on the Brønsted acid additive used. These include supported metal catalysts containing tungsten oxide or heteropolyacid or Rhenium as a cocatalyst/ Brønsted acid additive. All the supported metal catalysts with different types of additives used for 1,3-PDO are herein summarized systematically.

#### *3.3.2.1 Metal-tungsten-based catalytic materials*

A number of supported metal catalysts comprising WOx as a cocatalyst were actively investigated in both liquid phase and vapor phase reaction processes of glycerol hydrogenolysis of glycerol and showed that platinum-tungsten-based catalysts were the most efficient catalytic materials for the selective production of 1,3-PDO. A tungstate consists of an oxoanion of tungsten, and the simplest tungstate ion is $WO_2^{-4}$. Tungsten in the presence of hydrogen undergoes reduction ($W^{6+}$ to $W^{5+}$) and produces acidic protons ($H^+$). Various examples of Pt-tungsten catalysts used in glycerol hydrogenolysis for selective 1,3-PDO formation are summarized in **Table 2**. The use of tungsten in glycerol hydrogenolysis started in 1980s by the innovative work led by Celanese Corporation using a homogeneous rhodium complex ($Rh(CO)_2(acac)$) + tungstic acid catalyst [21]. In comparison to a simple protic acid, the addition of $H_2WO_4$ to the catalyst enhanced the catalytic performance to propanediols. Later extensive research work was carried out based on noble metal- and tungsten-based catalysts (Pt-WOx) [24–42]. While Pt/WOx catalysts stand out to be the most outstanding high-performance catalysts for selective hydrogenolysis of glycerol to 1,3-PDO compared to other noble metals, Cu-based WOx supported on titania catalyst has also been investigated and was found to efficient in the production of 1,3-PDO [42]. Though the mechanism is still uncertain, it is obvious from previous research that the addition of tungsten as a cocatalyst has very well enhanced the acid strength especially the number of Brønsted acid sites which favors the formation of 1,3-PDO [11]. Therefore, the major conclusions from all the previous studies indicate that the Brønsted acid sites generated from tungstate species are responsible for dehydration of glycerol and the loaded platinum acts as active centre for the hydrogenation of the intermediate to 1,3-PDO.

#### *3.3.2.2 Metal-heteropolyacid-based catalytic materials*

Heteropolyacid-based metal-supported catalysts have also been investigated for selective hydrogenolysis of glycerol to 1,3-propanediol. In general, catalysis by heteropolyacids (HPAs) is a field of prominence, appealing growing attention throughout the world, in which many new and exciting developments are taking place in both research and technology. HPAs are polyoxometalates having metal-oxygen octahedra as the basic structural units. Among a wide variety of HPAs, those belonging to the so-called Keggin series such as silicotungstic acid and phosphotungstic acid are potential catalysts and can replace corrosive liquid acids to afford green chemical processes. HPAs possess strong Brønsted acidity as compared to other solid acid

| Author | Catalyst | Temp (°C) | $H_2$ (MPa) | Solvent | Process | *C* (%) | *S* (%) | Ref. |
|---|---|---|---|---|---|---|---|---|
| Kurosaka et al. 2008 | $Pt/WO_3/ZrO_2$ | 170 | 8 | DMI[a] | Liquid phase | 85.8 | 28.2 | [24] |
| Gong et al. 2010 | $Pt/WO_3/TiO_2/SiO_2$ | 180 | 5.5 | $H_2O$ | Liquid phase | 15.3 | 50.5 | [25] |
| Qin et al. 2010 | $Pt/WO_3/ZrO_2$ | 130 | 4 | — | Vapor phase | 70.2 | 45.6 | [26] |
| Liu et al. 2012 | $Pt/m\text{-}WO_3$ | 180 | 5.5 | — | Vapor phase | 18 | 39.2 | [27] |
| Dam et al. 2013 | $Pt/WOx/Al_2O_3$ | 200 | 4 | $H_2O$ | Liquid phase | 49 | 28 | [28] |
| Arundathi et al. 2013 | Pt/WOx/AlOOH | 180 | 5 | $H_2O$ | Liquid phase | 100 | 66 | [29] |
| Zhu et al. 2014 | $Pt/WO_3/ZrO_2/SiO_2$ | 180 | 5 | — | Vapor phase | 54.3 | 52.0 | [30] |
| Fernandez et al. 2015 | $Pt/WOx/Al_2O_3$ | 200 | 4.5 | $H_2O$ | Liquid phase | 53.1 | 51.9 | [31] |
| Zhu et al. 2015 | $Pt\text{-}WOx/Al_2O_3$ | 160 | 5 | — | Vapor phase | 64.2 | 66.1 | [32] |
| Priya et al. 2015 | $Pt\text{-}WO_3/SBA\text{-}15$ | 210 | 0.1 | — | Vapor phase | 86 | 42 | [33] |
| Wang et al. 2016 | Pt/WOx | 140 | 1 | $H_2O$ | Liquid phase | 2.2 | 45.7 | [34] |
| Ma et al. 2016 | Pt/ZrW | 180 | 8 | $H_2O$ | Liquid phase | 53.4 | 36 | [35] |
| Edake et al. 2017 | $Pt/WO_3/Al_2O_3$ | 260 | 0.1 | — | Vapor phase | 99 | 14 | [36] |
| Zhang et al. 2017 | AuPt/WOx | 140 | 1 | $H_2O$ | Liquid phase | 81.4 | 51.6 | [37] |
| Liu et al. 2017 | $Pt\text{-}WO_3\text{-}Al_2O_3\text{-}SiO_2$ | 160 | 6 | $H_2O$ | Liquid phase | 48 | 56 | [38] |
| Shi et al. 2018 | Pt-WOx/SAPO-34 | 210 | 6 | $H_2O$ | Liquid phase | 44.3 | 19.2 | [39] |
| Wang et al. 2018 | $Pt/Au/WO_3$ | 155 | 5 | $H_2O$ | Liquid phase | 30.7 | 54.3 | [40] |
| Chen et al. 2018 | $Pt\text{-}Li_2B_4O_7/WOx/ZrO_2$ | 150 | 4 | $H_2O$ | Liquid phase | 90.7 | 36 | [41] |
| Wu et al. 2018 | $Cu\text{-}WOx\text{-}TiO_2$ | 180 | 3.5 | $H_2O$ | Liquid phase | 12.7 | 32.3 | [42] |

[a]*DMI, 1,3-dimethyl-2-imidazolidinone; C, conversion of glycerol; S, selectivity of 1,3-PDO.*

**Table 2.**
*Metal-tungsten-based catalysts for glycerol hydrogenolysis into 1,3-propanediol.*

catalysts and have been widely employed in several homogeneous and heterogeneous catalytic reactions. Much research has been focussed on heterogeneous catalysts containing active metal and heteropolyacid as a cocatalyst supported on metal oxides for the glycerol hydrogenolysis to selectively produce 1,3-propanediol (**Table 3**). The results showed that the catalysts exhibited better activity in glycerol conversion and 1,3-PDO selectivity which is ascribed to the synergic interaction of Keggin structure of HPA with the active metal as well as stability of the catalyst. HPA-based catalysts

| Author | Catalyst | Temp (°C) | $H_2$ (MPa) | Solvent | Process | C (%) | S (%) | Ref. |
|---|---|---|---|---|---|---|---|---|
| Huang et al. 2009 | $Cu-H_4SiW_{12}O_{40}/SiO_2$ | 210 | 0.54 | — | Vapor phase | 83.4 | 32.1 | [43] |
| Zhu et al. 2012 | $Pt-H_4SiW_{12}O_{40}/SiO_2$ | 200 | 6 | — | Vapor phase | 81.2 | 31.4[a] | [3] |
| Zhu et al. 2013 | $Pt-H_4SiW_{12}O_{40}/ZrO_2$ | 180 | 5 | — | Vapor phase | 26.7 | 38.9 | [44] |
| Zhu et al. 2013 | $Pt-H_4SiW_{12}O_{40}/ZrO_2$ | 180 | 5 | — | Vapor phase | 24.1 | 48.1 | [45] |
| Shen et al. 2017 | $PtNps-HSiW/mAl_2O_3$ | 200 | 4 | $H_2O$ | Liquid phase | 60.5 | 33.3 | [46] |

*[a]C, conversion of glycerol; S, selectivity of 1,3-PDO; Yield.*

**Table 3.**
*Metal-heteropolyacid-based catalysts for glycerol hydrogenolysis into 1,3-PDO.*

provided an approach to tune the acidic property in terms of Brønsted acid sites in the catalyst and assisted in enhancing the selectivity of 1,3-PDO. Furthermore, the so far reported heteropolyacid-based platinum catalysts exhibited superior long-term performance because of the robust catalyst structure and good thermal stability.

### 3.3.2.3 Metal rhenium-based catalytic materials

Furthermore, rhenium-oxide-modified supported metal catalysts are proven to offer new opportunities for achieving high selectivity to 1,3-PDO. Rhenium exhibits similar chemical properties as that of tungsten in terms of metal oxide formation and reduction. Rhenium was found to be an effective cocatalyst for noble metal-based catalysts. In the recent past, iridium- and rhodium-based catalysts have been widely explored for selective production of 1, 3-PDO by glycerol hydrogenolysis. Ir-ReOx/$SiO_2$ catalyst was found to be the most effective one for hydrogenolysis of glycerol into 1,3-PDO. The catalytic activity of various M1-ReOx/$SiO_2$ (M1 = Pt, Pd, Ru, Ir and Rh) and Ir-$M_2$Ox/$SiO_2$ (M2 = Mo, Re, W, Cr, Ag and Mn) catalysts has been investigated and screened to conclude that Ir-ReOx/$SiO_2$ catalyst achieved the highest 1,3-PDO yield (38%) at 81% glycerol conversion. It was also observed that the method of catalyst preparation and the metal precursors had an effect on the catalytic performance. A list of all different rhenium-based catalysts screened for glycerol hydrogenolysis to 1,3-PDO is included in **Table 4**. The mechanism of glycerol hydrogenolysis to selectively produce 1,3-propanediol over Ir-ReOx-based catalysts is presented in **Figure 13**. Ir-ReOx is a solid acid catalyst with two active centres, Ir being the hydrogen activation centre and Re-OH being the substrate activation centre. Glycerol adsorbs on Re-OH sites to form alkoxide species with hydroxyl groups of glycerol. The activation of hydrogen to form a hydride species takes place in the interface between Ir and Re. The hydride species attacks the neighboring C—O bond by a nucleophilic substitution reaction ($SN^2$-type), and subsequent hydrolysis of the alkoxide produces 1,3-propanediol. The combined synergic interaction between positively charged Re, acting as a Brønsted acid site, and noble metal Ir with high hydrogenating capacity has strengthened the hydrogenolysis reaction and favored the formation of 1,3-propanediol.

### 3.3.2.4 Other supported metal catalysts

Several other supported metal catalysts have been extensively tested for liquid phase and vapor phase glycerol hydrogenolysis which included various

| Author | Catalyst | Temp (°C) | $H_2$ (MPa) | Solvent | Process | *C* (%) | *S* (%) | Ref. |
|---|---|---|---|---|---|---|---|---|
| Werpy et al. 2006 | Ni-Re/C | 230 | 9 | $H_2O$ | Liquid phase | 47 | 4.6 | [47] |
| | Pd-Re/C | | | | | 26 | 5.2 | |
| Shimao et al. 2009 | Rh-ReOx/$SiO_2$ | 160 | 8 | $H_2O$ | Liquid phase | 86 | 10.4 | [48] |
| Ma et al. 2009 | Ru-Re/$SiO_2$ | 160 | 8 | $H_2O$ | Liquid phase | 51.7 | 4.2 | [49] |
| | Ru-Re/$ZrO_2$ | | | | | 56.9 | 5.5 | |
| | Ru-Re/Hβ | | | | | 52.8 | 2.7 | |
| | Ru-Re/HZSM-5 | | | | | 54.2 | 2.9 | |
| | Ru-Re/$TiO_2$ | | | | | 36.3 | 8.1 | |
| Nakagava et al. 2010 | Ir-ReOx/$SiO_2$ | 120 | 8 | $H_2O$ | Liquid phase | 81 | 46 | [50] |
| Shinmi et al. 2010 | Rh-ReOx/$SiO_2$ | 120 | 8 | $H_2O$ | Liquid phase | 79 | 14 | [51] |
| Ma et al. 2010 | Ru-Re/$SiO_2$ | 160 | 8 | $H_2O$ | Liquid phase | 23.8 | 13.6 | [52] |
| Daniel et al. 2010 | Pt-Re/C | 170 | 4 | | | 45 | 30 | [16] |
| Yamada et al. 2011 | Ir-ReOx/$SiO_2$ | 120 | 8 | $H_2O$ | Liquid phase | 57.6 | 45.1 | [12] |
| Nakagava et al. 2012 | Ir-ReOx/ $SiO_2$ + HZSM-5 | 120 | 8 | $H_2O$ | Liquid phase | 58.8 | 44.7 | [17] |
| Deng et al. 2015 | Ir-Re/KIT-6[a] | 120 | 8 | $H_2O$ | Liquid phase | 46.3 | 54.1 | [53] |
| Falcone et al. 2015 | Pt-Re/$SiO_2$ | 120 | 4 | $H_2O$ | Liquid phase | 8.3 | 24 | [54] |
| | Pt-Re/$SiO_2$-W | | | | | 9.1 | 29 | |
| | Pt-Re/C | | | | | 6.5 | 34 | |
| Luo et al. 2016 | Egg shell type Ir-ReOx | 130 | 8 | — | Vapor phase | 48.9 | 37.0 | [55] |
| Salagre et al. 2017 | Ni-CuRe/acid mesoporous saponite | 120 | 5 | Glycidol+sulfolane | Liquid phase | 98 | 35 | [56] |

[a]*KIT-6, mesoporous silica; C, conversion of glycerol; S, selectivity of 1,3-PDO.*

**Table 4.**
*Metal rhenium-based catalysts for glycerol hydrogenolysis into 1,3-PDO.*

active support materials such as zeolites (HZSM-5, Hβ, Y-zeolite and mordenite), metal oxides (zirconia, titania, alumina and silica), activated carbon, aluminum phosphate, mordenite and montmorillonite. In these catalysts, the support itself provided Brønsted acid sites necessary for the dehydration of glycerol to 3-HPA. Besides metal oxides, zeolites are proven to be efficient catalysts for glycerol hydrogenolysis with a greater degree of product control as the reaction occurs within the pores of the zeolite. Zeolites are solid acid hydrated aluminosilicate catalysts with relatively large open pores and a regular three-dimensional crystal structure. Most of the researches focused on the use of zeolites in gas phase dehydration of glycerol to acrolein due to their versatility.

In our earlier research, we reported 2 wt% Pt supported on mordenite as a highly active and selective catalyst in the hydrogenolysis of glycerol with 48.6% 1,3-PDO selectivity at 94.9% glycerol conversion [57]. In our recent study, we have

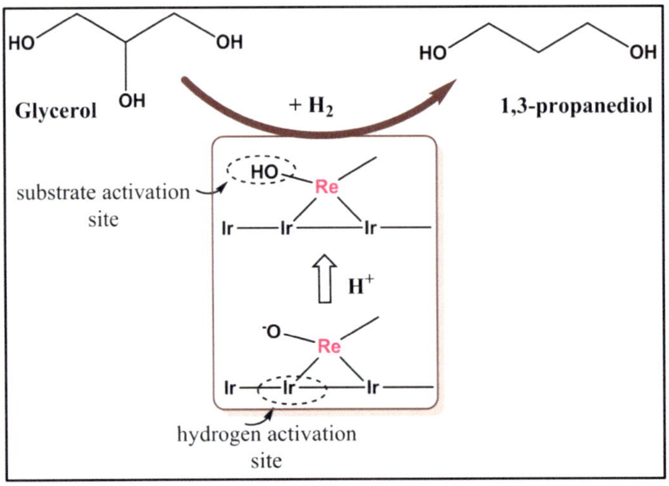


**Figure 13.**
*Selective hydrogenolysis of glycerol to 1,3-PDO over Ir-Re-based catalysts.*

| **Author** | **Catalyst** | **Temp (°C)** | **$H_2$ (MPa)** | **Solvent/Additive** | **Process** | ***C* (%)** | ***S* (%)** | **Ref.** |
|---|---|---|---|---|---|---|---|---|
| Chaminand et al. 2004 | 5% Rh/C | 180 | 8 | $H_2O/H_2WO_4$ | Liquid phase | 32 | 12 | [60] |
| | 1% Rh/Naflon | | | | | 8 | 19 | |
| Kusunoki et al. 2005 | Rh/C | 140 | 8 | $H_2O$/Amberlyst | Liquid phase | 6.4 | 7.2 | [61] |
| Miyazawa et al. 2006 | Ru/C | 120 | 8 | $H_2O$/Amberlyst | Liquid phase | 6.4 | 5.6 | [62] |
| Furikado et al. 2007 | Rh/$SiO_2$ | 120 | 8 | $H_2O$/Amberlyst | Liquid phase | 14 | 9.8 | [63] |
| Alhanash et al. 2008 | Rh/$Cs_{2.5}H_{0.5}$ [$PW_{12}O_{40}$] | 180 | 0.5 | $H_2O$ | Liquid phase | 6.3 | 7.1 | [64] |
| Ma et al. 2009 | Ru/$SiO_2$ | 160 | 8 | $H_2O$ | Liquid phase | 16.8 | 6.4 | [49] |
| | Ru/$ZrO_2$ | | | | | 25.4 | 1.8 | |
| | Ru/HZSM-5 | | | | | 20.5 | 6.0 | |
| | Ru/Hβ | | | | | 32.0 | 4.0 | |
| | Ru/$TiO_2$ | | | | | 6.0 | 6.8 | |
| Zhou et al. 2010 | Cu-Cr/$Al_2O_3$ | 200 | 1.5 | $H_2O$ | Liquid phase | 1 | 2 | [65] |
| | Cu-Cr/HY | | | | | 2 | 2 | |
| | Cu-Cr/Hβ | | | | | 2 | 1 | |
| Oh et al. 2011 | Pt/sulfated zirconia | 170 | 7.3 | DMI[a] | Liquid phase | 66.5 | 83.6 | [59] |
| Priya et al. 2014 | Pt/$ZrO_2$ | 230 | 0.1 | — | Vapor phase | 68 | 6.7 | [66] |
| | Pt/γ-$Al_2O_3$ | | | | | 55 | 2.8 | |
| | Pt/C | | | | | 86 | 5.8 | |
| | Pt/Y-zeolite | | | | | 100 | 4.9 | |
| Priya et al. 2014 | Pt/$AlPO_4$ | 260 | 0.1 | — | Vapor phase | 100 | 35.4 | [67] |
| Priya et al. 2016 | Pt/mordenite | 225 | 0.1 | — | Vapor phase | 94.9 | 48.6 | [57] |

| Author | Catalyst | Temp (°C) | $H_2$ (MPa) | Solvent/Additive | Process | *C* (%) | *S* (%) | Ref. |
|---|---|---|---|---|---|---|---|---|
| Priya et al. 2016 | Pt-Cu/mordenite | 210 | 0.1 | — | Vapor phase | 90 | 58.5 | [58] |
| Priya et al. 2018 | Pt/S-MMT | 200 | 0.1 | — | Vapor phase | 94 | 62 | [68] |

*[a]DMI, 1,3-dimethyl-2-imidazolidinone; C, conversion of glycerol; S, selectivity of 1,3-PDO.*

**Table 5.**
*Other supported metal catalysts for glycerol hydrogenolysis into 1,3-propanediol.*

successfully demonstrated that sulfuric acid-activated montmorillonite-supported platinum catalysts achieved highest selectivity (62%) to 1,3-PDO at 94% glycerol conversion under mild reaction conditions [58]. The modification of montmorillonite by acid treatment has significantly increased the acid catalytic performance of montmorillonite. Its combination with noble metal, platinum, which has been demonstrated to hold high hydrogenation activity and selectivity in C—O bond hydrogenolysis, was therefore a promising formulation to obtain high yields of 1,3-PDO possibly by two different mechanistic routes. Prior to this study, Oh et al. reported sulphated zirconia-supported platinum catalysts as super acid catalysts for selective formation of 1,3-PDO by glycerol hydrogenolysis [59] again based on the fact that Brønsted acid sites facilitated the selective production of 1,3-PDO. Furthermore, the mechanism of glycerol hydrogenolysis and the selective production of propanediols depending on the nature of acidic sites (Lewis or Brønsted) present in the catalyst have been clearly illustrated in another research based on platinum-copper bifunctional catalysts supported on mordenite [58]. Examples of other supported metal catalysts are summarized in **Table 5**.

## 4. Conclusions and future perspectives

Glycerol, one of the top 12 building block chemicals, can serve as a feedstock for the production of valuable fuel, fuel additives and chemical products. One of the most attractive route of upgrading glycerol is the formation of '1,3-propanediol' (1,3-PDO), a valuable industrial chemical with wide applications from carpet and textile manufacturing to cosmetics, personal and home care industry. Glycerol can be selectively converted to 1,3-PDO over a metal catalyst and $H_2$ via hydrogenolysis reaction. In spite of many past studies on selective glycerol hydrogenolysis, knowledge in order to achieve sustainable and economically competitive processes is still lacking and needs to be addressed. This need necessitates focusing research on developing highly efficient, economic viable and environmental-friendly catalytic processes. It is concluded that the supported metal catalysts with appropriate metal/acid balance were highly desirable for selective glycerol hydrogenolysis. Surface acidic sites as well as their strength in the support were found to be the important factors in controlling the product selectivity in glycerol hydrogenolysis. In addition to the acidic strength, the nature of acidic sites (Brønsted and Lewis) also played a key role in the product formation. Irrespective of the presence of weak, moderate or strong acidic sites in the support, Brønsted acidic sites selectively led to the formation of 1,3-propanediol, whereas Lewis acid sites favored the formation of 1,2-propanediol. In this chapter, selective glycerol hydrogenolysis to 1,3-PDO through different methods focussing on the catalytic conversion of glycerol has been presented. The reaction mechanisms elucidating the selective formation of propanediols and the role of acidity in product formation have been clearly illustrated. The importance of Brønsted acidity towards 1,3-PDO formation and the different classes of catalytic materials reported have been presented. Overall, the bifunctional metal-acid catalysts employed in the selective hydrogenolysis of glycerol to 1,3-PDO usually contained noble metals (Pt, Ir, Rh) in combination with Brønsted acid components (tungsten/heteropolyacid/rhenium).

The catalytic hydrogenolysis of glycerol is a promising route to transform glycerol into useful compounds; however, in order to achieve acceptable selectivities, additional important work should be carried out to gain further insights and it still remains a challenge. A systematic understanding of the principles behind converting glycerol into 1,3-propanediol paves the way for future work on developing novel catalytic systems and reaction processes. The selectivity to 1,3-PDO can be improved with the appropriate choice of the acidic support, and its interaction with the metal and further development of novel ways would be an important and challenging area of research. Vapor phase hydrogenolysis reactions carried out under solventless conditions and atmospheric pressure provide an economic viable and eco-friendly process. However, except for a few reactions, most of the research work in glycerol hydrogenolysis was performed in liquid phase using hazardous reaction conditions which makes the process incompatible. Therefore, much attention has to be focussed on the development of an economically viable commercial production of 1,3-propanediol from glycerol, with respect to both catalytic materials and the reaction processes.

## Acknowledgements

SSP duly acknowledges the Department of Chemical Engineering and the faculty of Engineering Discovery Seed Funding, Monash University.

## Conflict of interest

The author declares no conflict of interest.

## Author details

Shanthi Priya Samudrala
Department of Chemical Engineering, Monash University, Melbourne, Australia

*Address all correspondence to: priya.shanthipriya@monash.edu

## References

[1] Bridgwater AV. Review of fast pyrolysis and product upgrading. Biomass and Bioenergy. 2011:1-27

[2] Zhou CHC, Beltramini JN, Fan YX, Lu GQM. Chemoselective catalytic conversion of glycerol as a biorenewable source to valuable commodity chemicals. Chemical Society Reviews. 2008;**37**:527-549

[3] Huang L, Zhu Y, Zheng H, Ding G, Li Y. Direct conversion of glycerol into 1,3-propanediolover Cu-$H_4SiW_{12}O_{40}$/$SiO_2$ in vapor phase. Catalysis Letters. 2009;**131**:312-320

[4] Kenar JA. Glycerol as a platform chemical: Sweet opportunities on the horizon. Lipid Technology. 2007;**19**:249-253

[5] Werpy T, Petersen G. Top Value Added Chemicals from Biomass. USA: N.R.E.L. U. S. Department of Energy; 2004. Available online at http: www.osti.gov/bridge

[6] Zheng Y, Chen X, Shen Y. Commodity chemicals derived from glycerol, an important biorefinery feedstock. Chemical Reviews. 2008;**108**:5253-5277

[7] Priya SS, Kumar VP, Kantam ML, Bhargava SK, Periasamy S, Chary KVR. Metal–acid bifunctional catalysts for selective hydrogenolysis of glycerol under atmospheric pressure: A highly selective route to produce propanols. Applied Catalysis A: General. 2015;**498**:88-98

[8] Wolfson A, Litvak G, Dlugy C, Shotland Y, Tavor D. Employing crude glycerol from biodiesel production as an alternative green reaction medium. Industrial Crops and Products. 2009;**30**:78-81

[9] Behr A, Eilting J, Irawadi K, Leschinski J, Lindner F. Improved utilisation of renewable resources: New important derivatives of glycerol. Green Chemistry. 2008;**10**:13-30

[10] Drozdzynska A, Leja K, Czaczyk K. Biotechnological production of 1,3-propanediol from crude glycerol. Journal of Biotechnology, Computational Biology and Bionanotechnology. 2011;**92**:92-100

[11] Nakagawa Y, Tomishige K. Heterogeneous catalysis of the glycerol hydrogenolysis. Catalysis Science & Technology. 2011;**1**:179-190

[12] Amada Y, Shinmi Y, Koso S, Kubota T, Nakagawa Y, Tomishige K. Reaction mechanism of the glycerol hydrogenolysis to 1,3-propanediol over Ir–ReOx/SiO2 catalyst. Applied Catalysis, B: Environmental. 2011;**105**:117-127

[13] S Zhu S, Gao X, Zhu Y, Li Y. Promoting effect of WOx on selective hydrogenolysis of glycerol to 1,3-propanediol over bifunctional Pt–WOx/$Al_2O_3$ catalysts. Journal of Molecular Catalysis A. 2015;**398**:391-398

[14] Alhanash A, Kozhevnikova EF, Kozhevnikov IV. Gas-phase dehydration of glycerol to acrolein catalyzed by caesium heteropolysalt. Applied Catalysis A: General. 2010;**378**:11-18

[15] Kraus GA. Synthetic methods for the preparation of 1,3-propanediol. Clean: Soil, Air, Water. 2008;**36**:648-651

[16] Daniel OM, DeLaRiva A, Kunkes EL, Datye AK, Dumesic JA, Davis RJ. X-ray absorption spectroscopy of bimetallic Pt–Re catalysts for hydrogenolysis of glycerol to propanediols. ChemCatChem. 2010;**2**:1107-1114

[17] Nakagawa Y, Ning X, Amada Y, Tomishige K. Solid acid co-catalyst

for the hydrogenolysis of glycerol to 1,3-propanediol over Ir-ReOx/ $SiO_2$. Applied Catalysis A: General. 2012;**433-434**:128-134

[18] Yazdani SS, Gonzalez R. Anaerobic fermentation of glycerol: A path to economic viability for the biofuels industry. Current Opinion in Biotechnology. 2007;**18**:213-219

[19] Dasari MA, Kiatsimkul PP, Sutterlin WR, Suppes GJ. Low-pressure hydrogenolysis of glycerol to propylene glycol. Applied Catalysis A: General. 2005;**281**:225-231

[20] Yuan ZL, Wu P, Gao J, Lu XY, Hou ZY, Zheng XM. Pt/solid-base: A predominant catalyst for glycerol hydrogenolysis in a base-free aqueous solution. Catalysis Letters. 2009;**130**:261-265

[21] Che TM. Production of propanediols. Patent US 4 642 394 Celanese Corporation 1987

[22] Drent E, Jager WW, Hydrogenolysis of glycerol. Patent US 6080898 Shell Oil Co; 2000

[23] Schlaf M, Ghosh P, Fagan PJ, Hauptman E, Bullock RM. Metal-catalyzed selective deoxygenation of diols to alcohols. Angewandte Chemie International Edition. 2001;**40**:3887-3890

[24] Kurosaka T, Maruyama H, Naribayashi I, Sasaki Y. Production of 1,3-propanediol by hydrogenolysis of glycerolcatalyzed by Pt/ $WO_3$/ $ZrO_2$. Cataysis Communications. 2008;**9**:1360-1363

[25] Gong L, Lu Y, Ding Y, Lin R, Li J, Dong W, et al. Selective hydrogenolysis of glycerol to 1,3-propanediol over a Pt/ $WO_3$/ $TiO_2$/ $SiO_2$ catalyst in aqueous media. Applied Catalysis A: General. 2010;**390**:119-126

[26] Qin LZ, Song MJ, Chen CL. Aqueous-phase deoxygenation of glycerol to 1,3-propanediol over Pt/ $WO_3$/ $ZrO_2$ catalysts in a fixed-bed reactor. Green Chemistry. 2010;**12**:1466-1472

[27] Longjie L, Yanhua Z, Aiqin W, Tao Z. Mesoporous $WO_3$ supported Pt catalyst for hydrogenolysis of glycerol to 1,3-propanediol. Chinese Journal of Catalysis. 2012;**33**:1257-1261

[28] Dam Jt, Djanashvili K, Kapteijn F, Hanefeld U. Pt/ $Al_2O_3$ catalyzed 1,3-propanediol formation from glycerol using tungsten additives. ChemCatChem. 2013;**5**:497-505

[29] Arundhathi R, Mizugaki T, Mitsudome T, Jitsukawa K, Kaneda K. Highly selective hydrogenolysis of glycerol to 1,3-propanediol over a boehmite-supported platinum/ tungsten catalyst. ChemSusChem. 2013;**6**:1345-1347

[30] Zhu S, Gao X, Zhu Y, Cui J, Zheng H, Li Y. $SiO_2$ promoted Pt/WOx/ $ZrO_2$ catalysts for the selective hydrogenolysis of glycerol to 1,3-propanediol. Applied Catalysis B: Environmental. 2014;**158-159**:391-399

[31] Fernandez SG, Gandarias I, Requies J, Guemez MB, Bennici S, Auroux A, et al. New approaches to the Pt/WOx/ $Al_2O_3$ catalytic system behavior for the selective glycerol hydrogenolysis to 1,3-propanediol. Journal of Catalysis. 2015;**323**:65-75

[32] Zhu S, Gaob X, Zhu Y, Li Y. Promoting effect of WOx on selective hydrogenolysis of glycerol to 1,3-propanediol over bifunctional Pt–WOx/ $Al_2O_3$ catalysts. Journal of Molecular Catalysis A: Chemical. 2015;**398**:391-398

[33] Priya SS, Kumar VP, Kantam ML, Bhargava SK, Srikanth A, Chary KVR. High efficiency conversion of

glycerol to 1,3-propanediol using a novel platinum–tungsten catalyst supported on SBA-15. Industrial and Engineering Chemistry Research. 2015;**54**:9104-9115

[34] Wang J, Zhao X, Lei N, Li L, Zhang L, Xu S, et al. Hydrogenolysis of glycerol to 1,3-propanediol under low hydrogen pressure over WOx-supported single/pseudo-single atom Pt catalyst. ChemSusChem. 2016:9 1-9 8

[35] Zhou W, Zhao Y, Wang Y, Wang S, Ma X. Glycerol hydrogenolysis to 1,3-propanediol on tungstate/zirconia-supported platinum: Hydrogen spillover facilitated by Pt(111) formation. ChemCatChem. 2016;**8**:3663-3671

[36] Edake M, Dalil M, Mahboub MJD, Dubois JL, Patiencea GS. Catalytic glycerol hydrogenolysis to 1,3-propanediol in a gas–solid fluidized bed. RSC Advances. 2017;**7**:3853-3860

[37] Zhao X, Wang J, Yang M, Lei N, Li L, Hou B, et al. Selective hydrogenolysis of glycerol to 1,3-propanediol: Manipulating the frustrated Lewis pairs by introducing gold to Pt/WOx. ChemSusChem. 2017;**10**:819-824

[38] Feng S, Zhao B, Liu L, Dong J. Platinum supported on $WO_3$-doped aluminosilicate: A highly efficient catalyst for selective hydrogenolysis of glycerol to 1,3-propanediol. Industrial and Engineering Chemistry Research. 2017;**56**:11065-11074

[39] Shi g, Xu J, Song Z, Cao Z, Jin K, Xu S, et al. Selective hydrogenolysis of glycerol to 1,3-propanediol over Pt-WOx/SAPO-34 catalysts. Molecular Catalysis. 2018;**456**:22-30

[40] Yang C, Zhang F, Lei N, Yang M, Liu F, Miao Z, et al. Understanding the promotional effect of Au on Pt/ $WO_3$ in hydrogenolysis of glycerol to 1,3-propanediol. Chinese Journal of Catalysis. 2018;**39**:1366-1372

[41] Zhu M, Chen C. Hydrogenolysis of glycerol to 1,3-propanediol over $Li_2B_4O_7$-modified tungsten–zirconium composite oxides supported platinum catalyst. Reaction Kinetics, Mechanisms and Catalysis. 2018;**124**:683-699. DOI: 10.1007/s11144-018-1379-z

[42] Li D, Zhou Z, Qin J, Li Y, Liu Z, Wu W. Cu-WOx-$TiO_2$ catalysts by modified evaporation-induced self-assembly method for glycerol hydrogenolysis to 1,3-propanediol. ChemistrySelect. 2018;**3**:2479-2486

[43] Toshio O. New catalytic functions of heteropoly compounds as solid acids. Catalysis Today. 2002;**73**:167-176

[44] Zhu S, Zhu Y, Hao S, Chen L, Zhang B, Li Y. Aqueous-phase hydrogenolysis of glycerol to 1,3-propanediol over Pt-$H_4SiW_{12}O_{40}/SiO_2$. Catalysis Letters. 2012;**142**:267-274

[45] Zhu S, Qiu Y, Zhu Y, Hao S, Zheng H, Li Y. Hydrogenolysis of glycerol to 1,3-propanediol over bifunctional catalysts containing Pt and heteropolyacids. Catalysis Today. 2013;**212**:120-126

[46] Gu M, Shen Z, Yang L, Peng B, Dong W, Zhang W, et al. The effect of catalytic structure modification on hydrogenolysis of glycerol into 1,3-propanediol over platinum nanoparticles and ordered mesoporous alumina assembled catalysts. Industrial and Engineering Chemistry Research. 2017;**56**:13572-13581

[47] Werpy TA, Frye J, Zacher AH, Miller DJ. US 7038094 2006

[48] Shimao A, Koso S, Ueda N, Shinmi Y, Furikado I, Tomishige K. Promoting effect of Re addition to $Rh/SiO_2$ on glycerol Hydrogenolysis. Chemistry Letters. 2009;**38**:540-541

[49] Ma L, He D. Hydrogenolysis of glycerol to propanediols over highly

active Ru-Re bimetallic catalysts. Topics in Catalysis. 2009;**52**:834-844

[50] Nakagawa Y, Shinmi Y, Koso S, Tomishige K. Direct hydrogenolysis of glycerol to 1,3-propanediol over rhenium-modified iridium catalyst. Journal of Catalysis. 2010;**272**:191-194

[51] Shinmi Y, Koso S, Kubota T, Nakagawa Y, Tomishige K. Modification of Rh/$SiO_2$ catalyst for hydrogenolysis of glycerol in water. Applied Catalysis, B: Environmental. 2010;**94**:318-326

[52] Ma L, Li Y, He D. Glycerol hydrogenolysis to propanediols over Ru-Re/$SiO_2$: Acidity of catalyst and role of Re. Chinese Journal of Catalysis. 2011;**32**:872-876

[53] Deng C, Duan X, Zhou J, Zhou X, Yuan W, Scott SL. Ir-Re alloy as a highly active catalyst for the hydrogenolysis of glycerol to 1, 3-propanediol. Catalysis Science & Technology. 2015;**5**:1540-1547

[54] Falcone DD, Hack JH, Yu AK, Gericke AK, Schlögl R, Davis RJ. Evidence for the bifunctional nature of Pt-Re catalysts for selective glycerol hydrogenolysis. ACS Catalysis. 2015;**5**:5679-5695

[55] Luo W, Lyu Y, Gong L, Du H, Wang T, Ding Y. Selective hydrogenolysis of glycerol to 1,3- propanediol over egg-shell type Ir-ReOx catalysts. RSC Advances. 2016;**6**:13600-13608

[56] Gebrestadik FB, Llorca J, Salagre P, Cesteros Y. Hydrogenolysis of glycidol as alternative route to obtain selectively 1,3-propanediol using MOx modified Ni-Cu catalysts supported on acid mesoporous saponite. ChemCatChem. 2017;**9**:3670-3680

[57] Priya SS, Bhanuchander P, Kumar VP, Dumbre DK, Periasamy SR, Bhargava SK, et al. Platinum supported on H-mordenite: A highly efficient catalyst for selective Hydrogenolysis of glycerol to 1,3-propanediol. ACS Sustainable Chemistry & Engineering. 2016;**4**:1212-1222

[58] Priya SS, Bhanuchander P, Kumar VP, Bhargava SK, Chary KVR. Activity and selectivity of platinum–copper bimetallic catalysts supported on mordenite for glycerol hydrogenolysis to 1,3-propanediol. Industrial and Engineering Chemistry Research. 2016;**55**:4461-4472

[59] Oh J, Dash S, Lee H. Selective conversion of glycerol to 1,3-propanediol using Pt-sulfated zirconia. Green Chemistry. 2011;**13**:2004-2007

[60] Chaminand J, Djakovitch L, Gallezot P, Marion P, Pinel C, Rosier C. Glycerol hydrogenolysis on heterogeneous catalysts. Green Chemistry. 2004;**6**:359-361

[61] Kusunoki Y, Miyazawa T, Kunimori K, Tomishige K. Highly active metal–acid bifunctional catalyst system for hydrogenolysis of glycerol under mild reaction conditions. Catalysis Communications. 2005;**6**:645-649

[62] Miyazawa T, Kusunoki Y, Kunimori K, Tomishige K. Glycerol conversion in the aqueous solution under hydrogen over Ru/C+ an ion-exchange resin and its reaction mechanism. Journal of Catalysis. 2006;**240**:213-221

[63] Furikado I, Miyazawa T, Koso S, Shimao A, Kunimoria K, Tomishige K. Catalytic performance of Rh/$SiO_2$ in glycerol reaction under hydrogen. Green Chemistry. 2007;**9**:582-588

[64] Alhanash A, Kozhevnikova EF, Kozhevnikov IV. Hydrogenolysis of glycerol to propanediol over Ru: Polyoxometalate bifunctional catalyst. Catalysis Letters. 2008;**120**:307-311

[65] Zhou J, Guo L, Guo X, Mao J, Zhang S. Selective hydrogenolysis of glycerol to propanediols on supported

Cu-containing bimetallic catalysts. Green Chemistry. 2010;**12**:1835-1843

[66] Priya SS, Kumar VP, Kantam ML, Bhargava SK, Chary KVR. Vapour-phase hydrogenolysis of glycerol to 1,3-propanediol over supported Pt catalysts: The effect of supports on the catalytic functionalities. Catalysis Letters. 2014;**144**:2129-2143

[67] Priya SS, Kumar VP, Kantam ML, Bhargava SK, Chary KVR. Catalytic performance of Pt/$AlPO_4$ catalysts for selective hydrogenolysis of glycerol to 1,3-propanediol in the vapour phase. RSC Advances. 2014;**4**:51893-51903

[68] Priya SS, Kandasamy S, Bhattacharya S. Turning biodiesel waste glycerol into 1,3-propanediol: Catalytic performance of sulphuric acid-activated montmorillonite supported platinum catalysts in glycerol hydrogenolysis. Scientific Reports. 2018;**8**:7484

Chapter 6

# Glycerol as a Superior Electron Source in Sacrificial $H_2$ Production over $TiO_2$ Photocatalyst

*Masahide Yasuda, Tomoko Matsumoto and Toshiaki Yamashita*

## Abstract

Biodiesel fuel (BDF) has gained much attention as a new sustainable energy alternative to petroleum-based fuels. BDF is produced by transesterification of vegetable oil or animal fats with methanol along with the co-production of glycerol. Indeed, transesterification of vegetable oil (136.5 g) with methanol (23.8 g) was performed under heating at 61°C for 2 h in the presence of NaOH (0.485 g) to produce methyl alkanoate (BDF) and glycerol in 83.7 and 73.3% yields, respectively. Although BDF was easily isolated by phase separation from the reaction mixture, glycerol and unreacted methanol remained as waste. In order to construct a clean BDF synthesis, the aqueous solution of glycerol and methanol was subjected to sacrificial $H_2$ production over a Pt-loaded $TiO_2$ catalyst under UV irradiation by high-pressure mercury lamp. $H_2$ was produced in high yield. The combustion energy ($\Delta H$) of the evolved $H_2$ reached 100.7% of the total $\Delta H$ of glycerol and methanol. Thus, sacrificial agents such as glycerol and methanol with all of the carbon attached to oxygen atoms can continue to serve as an electron source until their sacrificial ability was exhausted. Sacrificial $H_2$ production will provide a promising approach in the utilization of by-products derived from BDF synthesis.

**Keywords:** BDF, photocatalyst, $TiO_2$, sacrificial agent, glycerol, hydrogen

## 1. Introduction

The major issue in the current world is an urgent need to stop the increase of $CO_2$ levels. A large amount of consumption of fossil resources causes serious environmental problems such as global warming and air pollution. Therefore, biofuels such as bioethanol, bio-hydrogen, and biodiesel (BDF) have gained much attention as renewable and sustainable energy alternative to petroleum-based fuels [1]. However, the problems to be solved for practical uses still remain in each biofuel. In bioethanol, the ethanol concentrations are still too low to isolate pure ethanol by distillation at a low energy cost [2, 3]. Bio-hydrogen is isolated spontaneously from reaction mixtures without operations to separate. However, it is needed to construct newly a supply system to vehicles.

BDF is produced by transesterification of vegetable oil or animal fats with methanol along with the co-production of glycerol [Eq. (1)] [4]. Although methyl alkanoate (BDF) is easily isolated by phase separation, a mixture of glycerol and unreacted methanol remains in aqueous solution as waste. New utilization of these

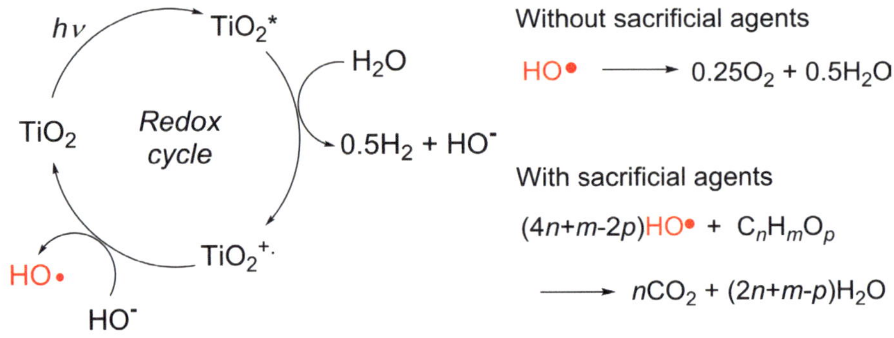


**Figure 1.**
*Photocatalytic water splitting over $TiO_2$.*

wastes is required. Reforming of glycerol has been extensively investigated through pyrolysis [5, 6], steam gasification [7, 8], and biological reforming [9, 10]. We have focused on photocatalytic reforming over titanium dioxide ($TiO_2$) [11]:

$$\begin{array}{l} RCO_2{-}CH_2 \\ RCO_2{-}CH \\ RCO_2{-}CH_2 \end{array} + 3CH_3OH \xrightarrow[NaOH]{} 3RCO_2CH_3 + \begin{array}{l} HO{-}CH_2 \\ HO{-}CH \\ HO{-}CH_2 \end{array} \qquad (1)$$

$TiO_2$ has a semiconductor structure with 3.2 eV of bandgap, which corresponds to 385 nm of light wavelength [12]. Therefore, the $TiO_2$ can be excited by 366 nm emitted from a high-pressure mercury lamp. Irradiation of the $TiO_2$ induces charge separation into electrons and holes (**Figure 1**). Electron excited to the conduction band serves to reduce water to $H_2$. Evolution of $H_2$ is usually accelerated by deposition of noble metals (Pt, Pd, and Au) onto the $TiO_2$. The positive charge (hole) oxidizes hydroxide absorbed on the surface of $TiO_2$ to generate hydroxyl radicals, which is eventually transformed to $O_2$ [13]. However, spontaneous conversion of hydroxyl radical into $O_2$ is inefficient. Moreover, water splitting into $O_2$ and $H_2$ is a large uphill reaction, resulting in rapid reverse reaction.

On the other hand, the hydroxyl radicals can be effectively consumed by the use of electron-donating sacrificial agents (hole scavengers), thus accelerating the $H_2$ production (**Figure 1**) [14]. This method is named "sacrificial $H_2$ production." The sacrificial $H_2$ production is an uphill process, but the energy change is small. Therefore, the sacrificial $H_2$ production proceeds more smoothly compared with water splitting without sacrificial agents, thus providing a convenient method to generate $H_2$ [15]. When one equivalent of hydroxyl radical is consumed, one equivalent of electron is generated to produce $0.5H_2$.

During our investigations on sacrificial $H_2$ production over a Pt-loaded $TiO_2$ (Pt/$TiO_2$) [15], it was found that sacrificial agents with all of the carbon attached oxygen atoms such as saccharides, polyalcohols (e.g., arabitol, glycerol, 1,2-ethandiol), and methanol continued to serve as an electron source until their sacrificial ability was exhausted. Glycerol (**1a**) and methanol (**1b**) are by-products from BDF synthesis. The **1a** has the potential to produce hydrogen in theoretical yield of seven equivalents, whose combustion energy ($\Delta H$ = 1995 kJ mol$^{-1}$) is larger than $\Delta H$ of **1a** (1654.3 kJ mol$^{-1}$) [Eq. (2)]. Also, **1b** can produce three equivalents of hydrogen, whose $\Delta H$ (855 kJ mol$^{-1}$) is larger than $\Delta H$ of **1b** (725.7 kJ mol$^{-1}$) [Eq. (3)] [16]. Thus, photo-energy can promote uphill process:

$$\underset{\mathbf{1a}\ (1654.3\ kJ\ mol^{-1})}{C_3H_8O_3 + 3H_2O} \xrightarrow[Pt/TiO_2]{h\nu} \underset{(1995\ kJ\ mol^{-1})}{3CO_2 + 7H_2} \qquad (2)$$

$$\underset{\mathbf{1b}\ (725.7\ \mathrm{kJ\ mol^{-1}})}{CH_3OH + H_2O} \xrightarrow[Pt/TiO_2]{h\nu} \underset{(855\ \mathrm{kJ\ mol^{-1}})}{CO_2 + 3H_2} \quad (3)$$

## 2. Outline of conversion of glycerol to hydrogen

Generally, biomass reforming is started by the production of water-soluble materials from biomass through biological treatment as well as chemical reaction [17, 18]. The resulting water-soluble materials (saccharides, amino acids) are converted to biofuels such as ethanol, methane, and hydrogen through various catalytic reactions in aqueous solution. Our biomass reforming is performed in aqueous solution through sacrificial $H_2$ production over $Pt/TiO_2$ using water-soluble materials derived from lignocelluloses [19–21] and chlorella [22] (**Figure 2**).

In this chapter, we will show $H_2$ production through sacrificial $H_2$ production over $Pt/TiO_2$ using **1a** and **1b** from standpoints of construction of renewable energy system and clean synthesis of BDF.

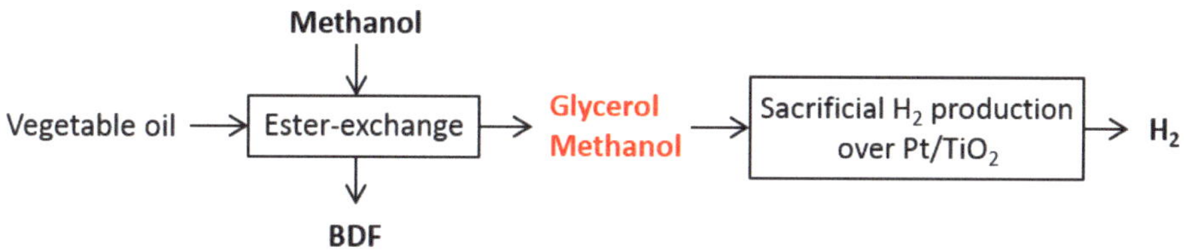


**Figure 2.**
*Outline of conversion of glycerol to hydrogen.*

## 3. Materials and method

### 3.1 Apparatus

NMR spectra were taken on a Bruker AV 400M spectrometer for $CDCl_3$ solution. LC-MS analysis were performed on a Waters Alliance 2695 under conditions (ESI ionization, capillary voltage 3.5 kV, source temperature 120°C and desolvation temperature 350°C) using column (Waters, SunFire C18, 2.1 mmΦ × 150 mm) and 1% formic acid in MeOH-$H_2O$ (6:4) as an eluent solution. GLC analysis of solution was performed on a Shimadzu 14A gas liquid chromatograph with FID detector at a temperature raised from 50 to 250°C using a capillary column (J & W CP-Sil 5CB, 0.32 mmΦ × 50 m).

### 3.2 Photoreaction apparatus

Reaction vessel was a cylindrical flask with 30 cm of height and 7.5 cm of diameter, which had three necks on the top. A high-pressure mercury lamp (100 W, UVL-100HA, Riko, Japan), which emitted mainly a light at 313 and 366 nm, was inserted into the large central neck of the reaction vessel. The reaction vessel was connected to a measuring cylinder with a gas-impermeable rubber tube to collect the evolved gas. The reaction vessel was set in a water bath to keep it at 20°C. The stirring of the solution was performed by magnetic stirrer.

### 3.3 Preparation of photocatalyst

Almost all research has used $TiO_2$ in anatase form such as P25 (Degussa Co. Ltd., Germany) and ST01 (Ishihara Sangyo Co. Ltd., Japan) for photocatalytic $H_2$ production. A Pt-loaded $TiO_2$ catalyst ($Pt/TiO_2$) was prepared by photo-deposition method according to the previous literature [23]. An aqueous solution (400 mL) containing $TiO_2$ (4.0 g, ST01), $K_2PtCl_6$ (40–400 mg), and 2-propanol (3.06 mL) was introduced reaction vessel, which was large scale of cylindrical flask with 35 cm of height and 9.0 cm of diameter. After the oxygen was purged by $N_2$ gas bubbling for 20 min, the solution was irradiated by stirring. After irradiation for 24 h, the water was entirely removed from the reaction mixture by an evaporator. The resulting black precipitate was washed with water on a filter and then dried under reduced pressure to produce $Pt/TiO_2$ [14]. The Pt content on $TiO_2$ was optimized to be 2.0 wt% by the comparison of the $H_2$ amounts evolved from photocatalytic reaction using **1a** (115 mg, 1.25 mmol) over various Pt contents of the Pt-doped $TiO_2$ (100 mg, 1.25 mmol) [15]. The structure of $Pt/TiO_2$ was analyzed by a Shimadzu XRD 7000 diffractometer.

### 3.4 Photocatalytic $H_2$ production

$Pt/TiO_2$ (100 mg) and the given amounts of aqueous solution of sacrificial agent were introduced to reaction vessel. The volume of the reaction solution was adjusted to 150 mL with water. Oxygen was purged from reaction vessel by $N_2$ gas for 20 min. $TiO_2$ was suspended in aqueous solution by vigorous stirring during the irradiation. Total volume of the evolved gas was measured by a measuring cylinder. Irradiation was performed until the gas evolution ceased. The evolved gas (0.5 mL) was taken through rubber tube using syringe and was subjected to the quantitative analysis of $H_2$, $N_2$, $CH_4$, and $CO_2$. Gas analysis was performed on a Shimadzu GC-8A equipped with TCD detector at temperature raised from 40 to 180°C using a stainless column (3 mmΦ, 6 m) packed with a SHINCARBON ST (Shimadzu).

In order to determine the quantum yield ($\Phi$) for $H_2$ evolution, the $H_2$ amount per hour was measured for various concentrations of **1** (8–40 mM). The $H_2$ amount per hour was converted to $\Phi$ using an actinometer which was $H_2$ amount per hour evolved from the sacrificial $H_2$ production using ethanol (0.434 M) at pH 10.0 over $Pt/TiO_2$ (Pt content 1.0 wt%), whose $\Phi$ was reported to be 0.057 [24]. Limiting quantum yields ($\Phi^{\infty}$) at an infinite concentration of **1** was determined from the intercept of the double reciprocal plots of $\Phi$ vs. the concentration of **1** [25].

## 4. Results

### 4.1 Sacrificial $H_2$ production using glycerol (1a) and methanol (1b)

Sacrificial $H_2$ production was applied to **1a** and **1b**. The $Pt/TiO_2$ (100 mg, 1.25 mmol, 2.0 wt% of Pt) was suspended in an aqueous solution (150 ml) of **1a** and **1b**, whose concentration was varied in a range of 0.25–1.25 mmol. After $O_2$ was purged from the reaction vessel using $N_2$ gas, UV irradiation was continued under vigorous stirring for 10–17 h until gas evolution had ceased [15]. The evolved gas volumes were plotted against the amounts of sacrificial agent used. In the absence of sacrificial agents, the evolved $H_2$ from water was small (<2 mL). **Figure 3A** is a typical example of the plots of volume of $H_2$ and $CO_2$ against the amounts of **1a** used. Gas volume increased as an increase of the

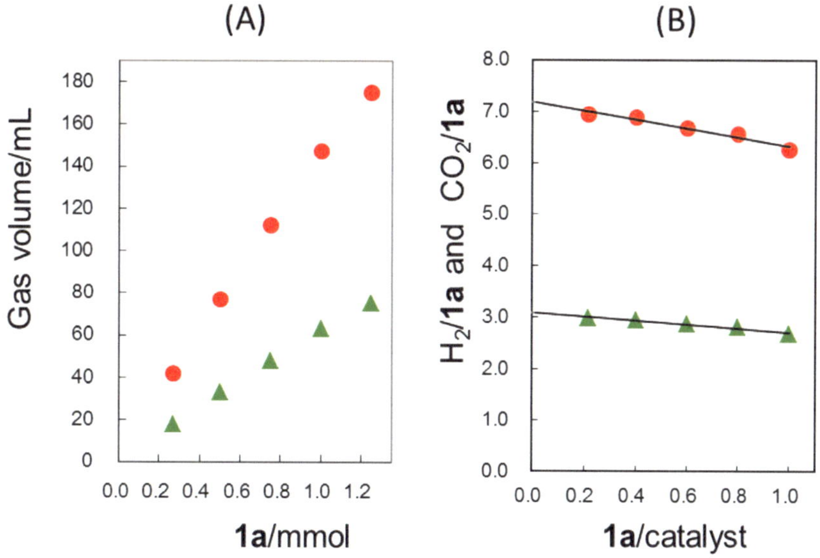


**Figure 3.**
*(A) The gas volume evolved from the sacrificial $H_2$ production using glycerol (1a) over Pt/$TiO_2$. (B) Plots of $H_2$/1a and $CO_2$/1a against 1a/catalyst: $H_2$ (●) and $CO_2$ (▲).*

amounts of **1a** used. However, the molar ratio of the evolved $H_2$ to **1a** ($H_2$/**1a**) was dependent on the amount of **1a** used. Therefore, the $H_2$/**1a** values were plotted against the molar ratios of **1a** to catalyst (**1a**/catalyst). This plot gave a good linear relationship, as shown in **Figure 3B**.

The intercept of the plot equaled the limiting amount of $H_2$ ($H_2^{max}$) obtained from 1 mol of **1a** when the amount of the catalyst was extrapolated to infinite. The $H_2^{max}$ became 7.2. The limiting amount of $CO_2$ ($CO_2^{max}$) obtained from 1 mol of **1a** at an infinite amount of the catalyst was also determined to be 3.1 from the plots of $CO_2$/**1a** against **1a**/catalyst (**Figure 3B**). The $H_2^{max}$ and $CO_2^{max}$ are summarized in **Table 1**. If the sacrificial agent ($C_nH_mO_p$) is entirely decomposed into $CO_2$ and $H_2O$ by hydroxyl radicals, theoretically $(2n + 0.5m - p)$ equivalents ($P$) of $H_2$ will be evolved in the $TiO_2$ photocatalytic reaction [Eq. (4)]. The $P$ values are listed in **Table 1**. Therefore, the chemical yield of $H_2$ production was defined to be 100 $H_2^{max}/P$. In the case of **1a**, the yield of $H_2$ production was found to be 103%. Also, the $CO_2^{max}$ value was close to the theoretical value. Similarly, the $H_2^{max}$ and $CO_2^{max}$ values of **1b** were determined to be 3.0 and 1.0, respectively. This shows that **1a** and **1b** are superior sacrificial agents, which are completely decomposed into $CO_2$ and water by sacrificial $H_2$ production:

$$C_nH_mO_p + (2n-p)H_2O \xrightarrow[\text{Pt/TiO}_2]{h\nu} nCO_2 + (2n+0.5m-p)H_2 \qquad P = 2n+0.5m-p \tag{4}$$

## 4.2 Degradation mechanism of 1a and 1b

Generally, hydroxyl radical can abstract hydrogen atom more efficiently from the hydroxylated carbon rather than the non-hydroxylated carbon. Therefore, degradation of alcoholic sacrificial agents proceeds through hydrogen-atom abstraction by hydroxyl radical from the hydroxylated alkyl group [Eq. (5)] [15]. Hydroxyl radical reacts with the secondary alcohols to produce ketones, which does not undergo further degradation. The primary alcohols reacted with hydroxyl radical to produce aldehyde, which undergoes further oxidation to carboxylic acid

| Sacrificial agents | Formula | $P^a$ | Products/mol mol$^{-1}$ | | | Yield/% $^b$ | $\Phi^{\infty c}$ |
|---|---|---|---|---|---|---|---|
| | | | $H_2^{max}$ | $CO_2^{max}$ | $CH_4^{max}$ | | |
| Alcohols | | | | | | | |
| Glycerol (**1a**) | $C_3H_8O_3$ | 7 | 7.2 | 3.1 | | 103 | 0.078 |
| Methanol (**1b**) | $CH_4O$ | 3 | 3.0 | 1.0 | | 100 | 0.057 |
| 1-Hydroxy-2-propanone (**1c**) | $C_3H_6O_2$ | 7 | 4.9 | 2.5 | 0.30 | 87 | 0.045 |
| 1,2-Propanediol (**1d**) | $C_3H_8O_2$ | 8 | 4.8 | 1.0 | Trace | 60 | |
| 1,3-Propanediol (**1e**) | $C_3H_8O_2$ | 8 | 4.2 | 0.5 | | 53 | |
| 1-Propanol (**1f**) | $C_3H_8O$ | 9 | 4.1 | 1.0 | | 46 | 0.069 |
| 2-Propanol (**1g**) | $C_3H_8O$ | 9 | 1.3 | 0.0 | | 14 | |
| Carboxylic acids | | | | | | | |
| Glycolic acid (**2a**) | $C_2H_4O_3$ | 3 | 2.8 | 1.8 | | 93 | |
| Oxalic acid (**2b**) | $C_2H_2O_4$ | 1 | 1.0 | 2.0 | | 100 | |
| Formic acid (**2c**) | $CH_2O_2$ | 1 | 1.0 | 1.0 | | 100 | |
| Acetic acid (**2d**) | $C_2H_4O_2$ | 4 | 2.9 | 1.7 | 0.27 | 100 | |
| Pyruvic acid (**2e**) | $C_3H_4O_3$ | 5 | 3.9 | 2.7 | 0.30 | 102 | |
| Lactic acid (**2f**) | $C_3H_6O_3$ | 6 | 4.1 | 2.3 | 0.30 | 88 | |
| Malonic acid (**2g**) | $C_3H_4O_4$ | 4 | 2.6 | 2.7 | 0.31 | 96 | |
| Propanoic acid (**2h**) | $C_3H_6O_2$ | 7 | 2.3 | 1.0 | | 33 | |

[a]*Theoretical amount of hydrogen was calculated using Eq. (4).*
[b]*Total chemical yield of $H_2$ and $CH_4$ = 100 ($H_2^{max}$ + $4CH_4^{max}$)/P.*
[c]*Limiting quantum yield ($\Phi^{\infty}$) for $H_2$ evolution with infinite amounts of 1.*

**Table 1.**
*Sacrificial $H_2$ production over Pt/TiO$_2$ using alcohols (1) and carboxylic acids (2).*

[Eq. (6)]. Furthermore, H abstraction from carboxylic acid by hydroxyl radical induces decarboxylation from carboxylic acids through the formation of carboxyl radical ($RCO_2\cdot$) [Eq. (7)]. When hydroxyl group was substituted on α-position of carboxylic acid [X = OH in Eq. (7)], the decarboxylation took place more smoothly. Many researchers proposed that the decomposition of carboxylic acids is initiated by hole transfer to the carboxylic group rather than H abstraction by hydroxyl radicals [26–29]. Thus, the degradation of alcohols proceeds through the formation of carboxylic acids:

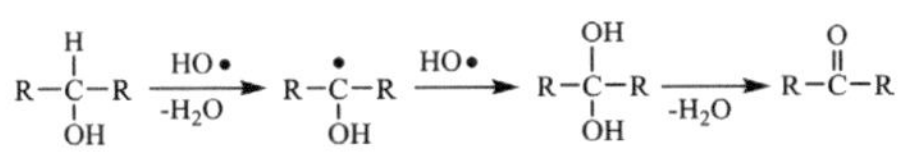
(5)

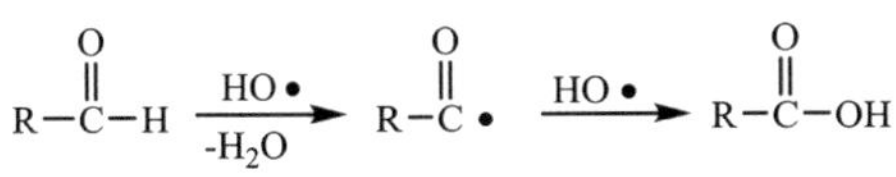
(6)

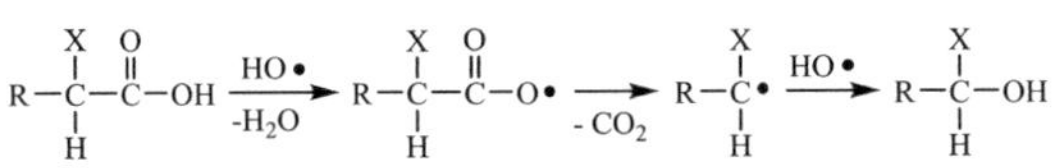
(7)

In 2009, Kondarides et al. reported sacrificial $H_2$ production from **1a** over Pt/TiO$_2$ (0.1–0.5 wt% Pt) [30]. They proposed that the decomposition of **1a** proceeded through the formation of methanol and acetic acid which were eventually decomposed into $CO_2$ and $H_2$ in a ratio of 3:7 [31]. Also, in irradiation of

$Pt/TiO_2$ in the absence and in the presence of glycerol, they detected $H_2O_2$ which was produced by dimerization of hydroxyl radicals [32]. Also, Ratnawati et al. detected a small amount of 1,2-ethanediol and acetic acid in reaction mixture [33]. They elucidated that Pt catalyzed not only reduction of water to $H_2$ but also dehydration of **1a**. Bowker et al. examined the photocatalytic reforming of **1a** over $M/TiO_2$ (M = 0.5 wt% Pd, 2.0 wt% Au) [34]. However, chemical yield of $H_2$ was still unclear.

We thought that degradation of **1a** was initiated by the oxidation of terminal alcohol by hydroxyl radical. It was thought that glycolic acid (**2a**) and oxalic acid (**2b**) were the intermediates intervening in degradation process of **1a**. Therefore, we performed sacrificial $H_2$ production over $Pt/TiO_2$ using **2a** and **2b**. The $H_2^{max}$ and $CO_2^{max}$ values of **2a** and **2b** were shown in **Table 1**. The **2a** and **2b** were completely decomposed to $CO_2$ and water, since the $CO_2^{max}$ values of **2a** and **2b** were determined to be 1.8 and 2.0, respectively. Although the degradation of **2a** could proceed through **2b** and/or formic acid (**2c**), we could not determine which degradation pathway occurred. In the case of **1b**, it was thought that **2c** was undoubtedly the intermediates intervening in degradation process of **1b**. The **2c** was completely decomposed to $CO_2$ and water, since the $CO_2^{max}$ value of **2c** was 1.0. However, **2a**, **2b**, and **2c** were not detected in the reaction mixture of sacrificial $H_2$ production using **1a** and **1b** due to easy decomposition of these carboxylic acids by hydroxyl radical. Also, Lu et al. have reported the degradation of **2b** and **2c**, which can adsorb on $Pt/TiO_2$ to give one equivalent $H_2$ under irradiation [35, 36].

According to Eqs. (5)–(7), a possible degradation mechanism of **1a** and **1b** by hydroxyl radical is shown in **Figure 4**. In the case of **1a**, 14 equivalents of hydroxyl radicals were consumed by **1a** along with the formation of $3CO_2$. At the same time, seven equivalents of $H_2$ were evolved. Actually, 7.2 of $H_2^{max}$ and 3.1 of $CO_2^{max}$ values of **1a** were provided from sacrificial $H_2$ production using **1a**. In the case of **1b**, six equivalents of hydroxyl radicals were consumed along with the formation of one equivalent of $CO_2$ and $3H_2$, providing actually 3.0 of $H_2^{max}$ and 1.0 of $CO_2^{max}$.

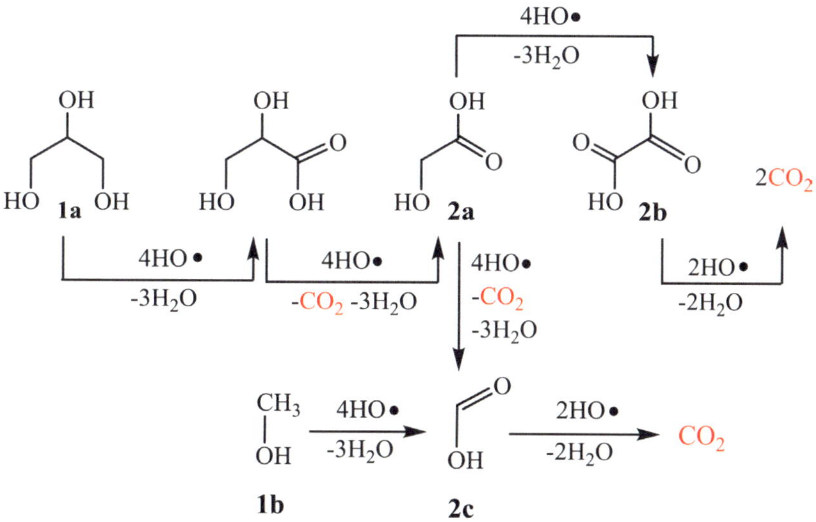


**Figure 4.**
*Degradation pathways of glycerol (**1a**) and methanol (**1b**) by hydroxyl radical in the sacrificial $H_2$ production over $Pt/TiO_2$.*

### 4.3 Structural dependence on $H_2$ yields in sacrificial $H_2$ production using several alcohols (1c–1g)

In order to elucidate the relationship between molecular structure of sacrificial agents and degradation yield, sacrificial $H_2$ production was performed using propane-based alcohols such as 1-hydroxy-2-propanone (**1c**); 1,2-propanediol (**1d**); 1,3-propanediol (**1e**); 1-popanol (**1f**); and 2-propanol (**1g**) (**Figure 5**) as well as the related carboxylic acids (**2d**–**2h**) [15].

Sacrificial $H_2$ evolution using **1c** produced $CH_4$ along with the formation of $H_2$ and $CO_2$. Limiting amount of $CH_4$ ($CH_4^{max}$) obtained from 1 mol of **1c** was 0.30 along with 4.9 of $H_2^{max}$ and 2.5 of $CO_2^{max}$ values. In the case of sacrificial $H_2$ production along with the formation of $CH_4$, the chemical yield was defined by the following equation: Yield = 100 $(H_2^{max} + 4CH_4^{max})/P$. The yield for the sacrificial $H_2$ production using **1c** was calculated to be 87%. Moreover, acetic acid (**2d**) was detected by LC-MS of the reaction solution at low conversion. A peak appeared at 2.24 min of retention time which showed mass peaks at $m/z$ 60 ($M^+$) and 43 ($CH_3CO^+$). Therefore, **2d** was subjected to sacrificial $H_2$ production. Mozia et al. reported that **2d** was decomposed into $H_2$, $CO_2$, and $CH_4$ over $TiO_2$ without Pt [37], although Zheng et al. reported that a trace amount of $CH_4$ was detected from **2d** over Pt/$TiO_2$ (Pt = 1.0 wt%) [38]. We determined the chemical yields [39]. The $CH_4^{max}$ of **2d** was determined to be 0.27 along with 2.9 of $H_2^{max}$ and 1.7 of $CO_2^{max}$ values. The total yield was calculated to be 100% (=100 (2.9 + 4 × 0.27)/4) in the sacrificial $H_2$ production using **2d**. Considering the experimental error, stoichiometric equation for conversion of **2d** into $H_2$, $CH_4$, and $CO_2$ was shown in Eq. (8):

$$\underset{\mathbf{2d}}{\mathrm{CH_3\overset{\overset{O}{\|}}{C}OH}} + 1.4\mathrm{H_2O} \xrightarrow[\mathrm{Pt/TiO_2}]{h\nu} 2.8\mathrm{H_2} + 1.7\mathrm{CO_2} + 0.3\mathrm{CH_4} \quad (8)$$

It was thought that pyruvic acid (**2e**) was an intermediate of degradation process from **1c** to **2c**. The $H_2^{max}$, $CO_2^{max}$, and $CH_4^{max}$ values of **2e** were found to be 3.9, 2.7, and 0.3, respectively [39]. Degradation scheme of **2e** can be expressed by Eq. (9). The yield for the sacrificial $H_2$ production using **2e** was 100%. Since the degradation yield of **1c** was found to be 87%, the degradation of **1c** to $H_2$, $CO_2$, and $CH_4$ proceeded effectively through the formation **2e** followed by **2d**:

$$\underset{\mathbf{2e}}{\mathrm{CH_3\overset{\overset{O}{\|}}{C}-\overset{\overset{O}{\|}}{C}OH}} + 2.4\mathrm{H_2O} \xrightarrow[\mathrm{Pt/TiO_2}]{h\nu} 3.8\mathrm{H_2} + 2.7\mathrm{CO_2} + 0.3\mathrm{CH_4} \quad (9)$$

The next sacrificial $H_2$ production was examined using **1d**. Oxidation of **1d** with hydroxyl radical was initiated by oxidation of primary alcohol part to afford lactic acid (**2f**). Sacrificial $H_2$ production using **2f** produced $H_2$, $CH_4$, and $CO_2$. The $H_2^{max}$, $CO_2^{max}$, and $CH_4^{max}$ values of **2f** were 4.1, 2.3, and 0.30, respectively. On the other hand, the $H_2^{max}$ and $CO_2^{max}$ values of **1d** were determined to be 4.8 and 1.0, respectively. Trace amount of $CH_4$ was formed. Thus, complete decomposition of **1d**

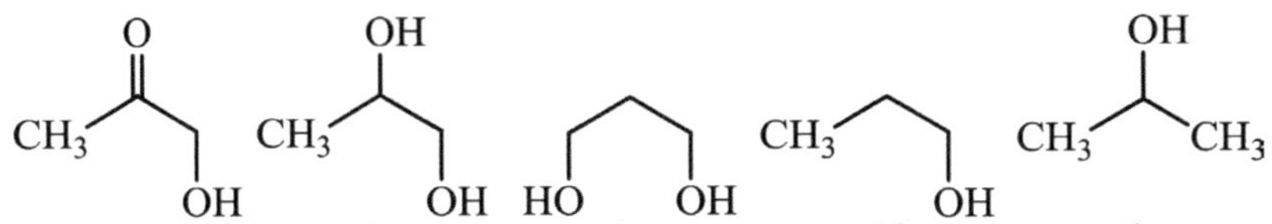

**1c** **1d** **1e** **1f** **1g**

**Figure 5.**
*Propane-based alcohols (1c–1g) as sacrificial agents for the photocatalytic $H_2$ production.*

into $H_2$ and $CO_2$ did not take place. Therefore, it is speculated that degradation of **1d** proceeds via **2f** which was decomposed to acetaldehyde. It is suggested that oxidation of acetaldehyde by hydroxyl radical was slow.

In sacrificial $H_2$ production using **1e**, $H_2^{max}$ and $CO_2^{max}$ values of **1e** were 4.2 and 0.50, respectively. Moreover, malonic acid (**2g**, *m/z* 104 ($M^+$)) was detected in LC-MS of the photolysate. The sacrificial $H_2$ production using **2g** showed that the $H_2^{max}$, $CO_2^{max}$, and $CH_4^{max}$ values were determined to be 2.6, 2.7, and 0.31, respectively. Degradation scheme of **2g** can be expressed by Eq. (10). Although the degradation yield of **2g** was relatively high yield (96%), **1e** was not completely decomposed, resulting in 0.5 of the $CO_2^{max}$ and no $CH_4$ emission. This suggests that the degradation process of **2g** is slow:

$$\underset{\textbf{2g}}{HO\overset{O}{\overset{\|}{C}}CH_2\overset{O}{\overset{\|}{C}}OH} + 1.4H_2O \xrightarrow[Pt/TiO_2]{h\nu} 2.8H_2 + 2.7CO_2 + 0.3CH_4 \tag{10}$$

Moreover, sacrificial $H_2$ production was applied to **1f**. The $H_2^{max}$ and $CO_2^{max}$ values of **1f** were determined to be 4.1 and 1.0, respectively. $CH_4$ was not formed. It is suggested that the degradation of **1f** proceeded via the formation of propanoic acid (**2h**). The $H_2^{max}$ and $CO_2^{max}$ values of **2h** were determined to be 2.3 and 1.0, respectively. The decarboxylation of **2h** and the subsequent oxidation gave acetaldehyde, which was subjected to the further degradation, but it was slow process [39]. In the case of sacrificial $H_2$ production using **1g**, acetone was detected by GLC analysis of the reaction mixture. The $H_2^{max}$ value was determined to be 1.3 and $CO_2$ was not evolved. Further degradation of acetone did not proceed.

Based on these results, the degradation pathways of **1c**, **1d**, **1e**, **1f**, and **1g** by hydroxyl radical are summarized in **Figure 6**. Though considerable amounts of $CO_2$

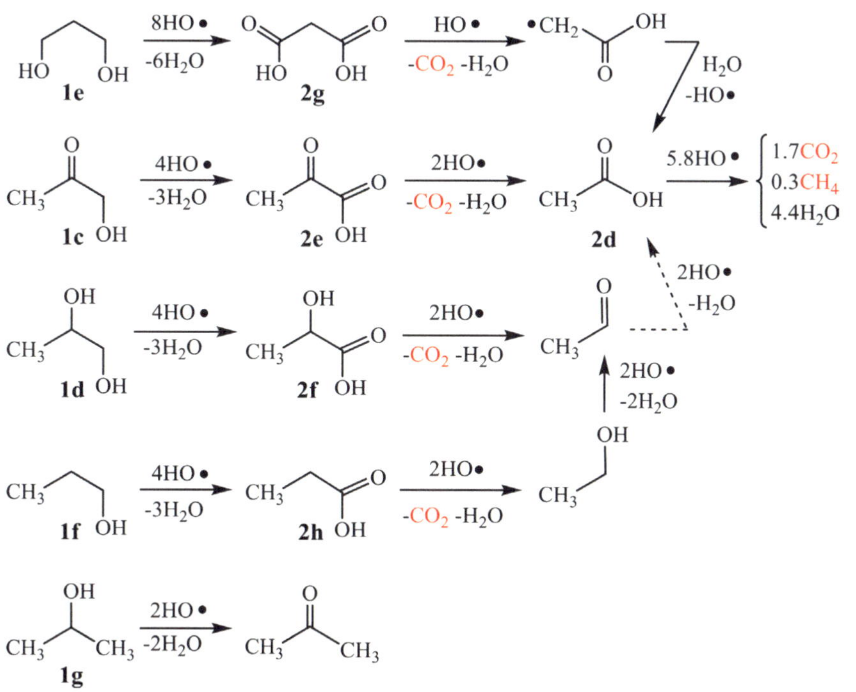


**Figure 6.**
*Degradation pathway of sacrificial agents by hydroxyl radical in the sacrificial $H_2$ production over Pt/$TiO_2$ using propane-based alcohols: 1-hydroxyl-2-propanone (1c); 1,2-propanediol (1d); 1,3-propanediol (1e); 1-propanol (1f); and 2-propanol (1g).*

were evolved from **1c** to **1f**, the $CO_2^{max}$ (0.5–2.5) did not reach the theoretical values. In the case of **1g**, $CO_2$ was not formed at all. Thus, in the case of these polyols which have one or two non-hydroxy-substituted carbons, the $H_2^{max}$ and $CO_2^{max}$ values did not reach the theoretical values. Therefore, we conclude that sacrificial agents with all of the carbon attached to oxygen atoms such as **1a** and **1b** continued to serve as an electron source until their sacrificial abilities were exhausted.

### 4.4 Separation of residual glycerol and methanol in BDF synthesis

Vegetable oil was mainly composed of the oleic acid ($C_{17}H_{33}CO_2H$) triglyceride whose average molecular weight was thought to be 884 g/mol. At first, since carboxylic acid was included in used oil as impurity, the amounts of NaOH (*a* g/kg-lipid) which was required to achieve pH of 8–9 were determined. Lipid (ca. 1 mL, 0.884 g) was solved in 2-propanol (10 mL) and neutralized by an aqueous NaOH solution. In this case, *a* was determined to be 0 g since fresh vegetable oil was used.

Vegetable oil (150 mL, 136.5 g, 0.154 mol) was set in a reaction vessel. Since usual optimal amount for transesterification of neutral lipid is known to be 3.55 g/kg [11], the amount of NaOH necessary to the transesterification was determined to be 0.485 g (=0 + 0.485 g) by the sum of *a* g/kg and 3.55 g/kg. Usually, 20% of weight of **1b** to vegetable oil is used for BDF synthesis. **1b** (30 mL, 23.8 g, 0.743 mol) was mixed with NaOH (0.485 g, 0.012 mol). About half of the mixture of **1b** and NaOH was poured in a reaction vessel and then kept at 61°C for 1 h under stirring. Moreover, the remaining

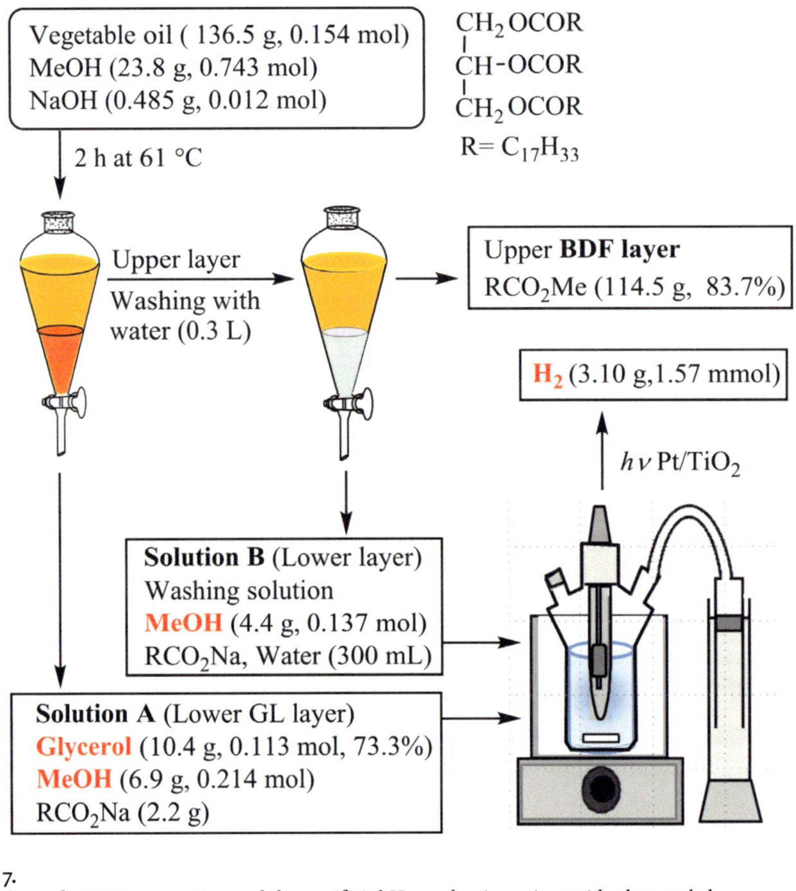


**Figure 7.**
*Mass balance for BDF preparation and the sacrificial $H_2$ production using residual 1a and 1b.*

mixture of **1b** and NaOH was added into the reaction vessel, and the reaction mixture was kept at 61°C for another 1 h.

Follow-up operation is shown in **Figure 7**. After cooling, the reaction mixture was separated into a lower layer and an upper layer. The lower layer (solution A) contained **1a** and **1b**. The upper layer was washed with water (300 mL) and separated to the BDF upper layer. Aqueous solution (solution B) was obtained from the lower layer. In order to check the contamination of lipid to BDF layer, the purity of BDF was determined by the peak-area ratio of methyl and methoxy groups in NMR spectra. The BDF layer contained $C_{17}H_{33}CO_2Me$ (114.5 g, 0.387 mol) and unreacted vegetable oil (2.2 g). The yield of $C_{17}H_{33}CO_2Me$ (BDF) was 83.7% based on the theoretical amounts of 137 g (0.463 mol).

GLC analysis showed that solution A contained **1a** (10.4 g, 0.113 mol) and **1b** (6.85 g, 0.214 mol) where molar ratio ($b$) of **1a** to **1b** was 1.89. The yield of **1a** was 73.3% based on the theoretical amounts of 14.2 g (0.154 mol). NMR analysis of solution A showed that $RCO_2Na$ (2.2 g) was contained in solution A. Solution B contained **1b** (4.38 g, 0.137 mol) and a small amount of $C_{17}H_{33}CO_2Na$. Thus, **1b** was found in both solutions A and B.

### 4.5 Hydrogen production from residual methanol and glycerol in BDF synthesis

The photocatalytic reforming of **1a** and **1b** was examined using solution A. Irradiation was performed by a high-pressure mercury lamp under vigorous stirring with a magnetic stirrer. **Figure 8** shows the plots of the $H_2$/**1a** against the molar ratio of **1a** to the catalyst (**1a**/catalyst), which was adjusted to 0.2, 0.4, 0.6, 0.8, and 1.0. From the intercept of the plots, $H_2^{max}$ obtained from 1 mol of **1a** at an infinite amount of the catalyst was determined to be 12.52. The yields of $H_2$ production of solution A were determined as follows. According to Eq. (11), the $H_2$ amount ($P$) was theoretically calculated to be 12.67 using $P = 7 + 3b$ and $b = 1.89$. Since actual $H_2^{max}$ was determined to be 12.52, the yield was calculated to be 98.8% (=$100H_2^{max}$/P). The results are summarized in **Table 2**:

$$C_3H_8O_3 + bCH_3OH + (3+b)H_2O \xrightarrow[Pt/TiO_2]{h\nu} (3+b)CO_2 + (7+3b)H_2 \quad P = 7 + 3b \tag{11}$$

$$CH_3OH + H_2O \xrightarrow[Pt/TiO_2]{h\nu} CO_2 + 3H_2 \quad P = 3 \tag{12}$$

Next, photocatalytic reforming was performed with solution B containing **1b**. Solution B was neutralized with dilute $H_2SO_4$ in order to reduce the effect of excess NaOH on $TiO_2$. After that, an aqueous solution (150 mL) containing **1b** (0.25–1.25 mmol) was irradiated in the presence of $Pt/TiO_2$ (100 mg) in a similar manner as solution A. The plots of $H_2$/**1b** against the molar ratio of **1b** to catalyst (**1b**/catalyst) are overlaid on **Figure 8**. The $H_2^{max}$ values were determined to be 1.08. The $H_2$ yields were calculated to be 36.0% based on the theoretical $P$ (3.00) [Eq. (12)]. In solution B, $C_{17}H_{33}CO_2Na$ was converted to $C_{17}H_{33}CO_2H$ by neutralization. It is well known that the carboxylic acid can strongly be adsorbed on $TiO_2$. Therefore, it is suggested that the adsorption of $C_{17}H_{33}CO_2H$ on $TiO_2$ lowered the photocatalytic activity of $TiO_2$. The presence of $C_{17}H_{33}CO_2H$ retarded the $H_2$ production of solution B remarkably.

Total amount of $H_2$ from solutions A and B was calculated to be 1.56 mol by Eq. (13) using 0.113 mol of **1a** in solution A and 0.137 mol of **1b** in solution B: $0.113 \times 12.52 + 0.137 \times 1.08$ (**Table 2**). $H_2$ (1.56 mol) whose combustion energy ($\Delta H$) was 445 kJ was evolved from solutions A and B. The $\Delta H$ of $H_2$ was compared with $\Delta H$

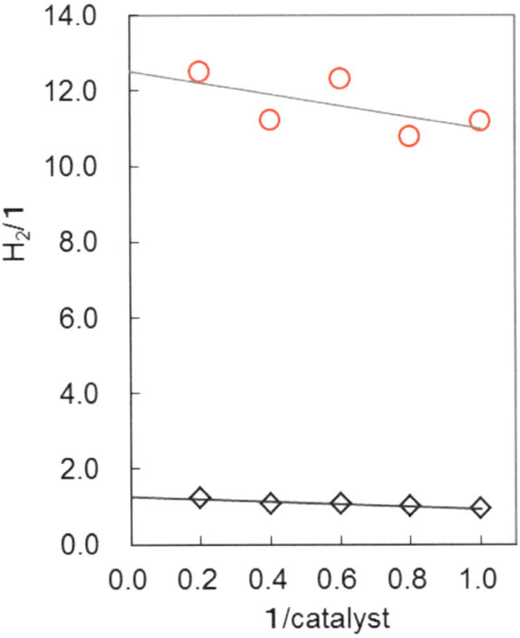


**Figure 8.**
*Determination of $H_2^{max}$ values by the plots of $H_2/$**1** against **1**/catalyst using solution A (○, b = 1.89) and solution B (◇) obtained from the BDF synthesis.*

| | Residues of BDF synthesis[a] | | | Photocatalytic reforming[b] | |
|---|---|---|---|---|---|
| | **1a/mol** | **1b/mol** | ***P***[c] | $H_2^{max}$ [d] | **$H_2$/mol (yield/%)**[e] |
| Solution A[f] | 0.113 | 0.214 | 12.67 | 12.52 | 1.41 (98.8) |
| Solution B | | 0.137 | 3.00 | 1.08 | 0.15 (36.0) |
| Total | 0.113 | 0.351 | | | 1.56 |
| [$\Delta H$/kJ][g] | [187] | [255] | | | [445] (100.7)[h] |

[a]*Transesterification was performed by the reaction of lipid (136.5 g, 0.154 mol) with* **1b** *(23.8 g, 0.743 mol) in the presence of NaOH (0.485 g, 0.012 mol) at 61°C for 2 h. BDF (114.5 g) was isolated.*
[b]*Photocatalytic reforming was performed by irradiation of Pt/$TiO_2$ in aqueous solution of* **1a** *and* **1b** *obtained from solutions A and B.*
[c]*The values were the theoretical amounts (P) obtained from Eqs. (11) and (12).*
[d]*The limiting amount of $H_2$ ($H_2^{max}$) was obtained from* **Figure 8**.
[e]*The values in parenthesis were the yield of $H_2 = 100H_2^{max}/P$.*
[f]*The molar ratio (b) of* **1b** *to* **1a** *was 1.89.*
[g]*The combustion energy ($\Delta H$) of* **1a**, **1b**, *and $H_2$ were 1654.3, 725.7, and 285.0 kJ $mol^{-1}$, respectively [16].*
[h]*The energy recovery yield was calculated to be 100.7% by Eq. (14).*

**Table 2.**
*Photocatalytic reforming of residues of BDF synthesis.*

of **1a** and **1b**. As shown in **Table 2**, 0.113 mol of **1a** and 0.351 mol of **1b** were isolated from BDF synthesis which had 442 kJ of $\Delta H$. The energy recovery yield was calculated to be 100.7% by using Eq. (14):

$$\text{Amount of } H_2 = (\text{mole of } \mathbf{1}) \times (H_2^{max}) \tag{13}$$

$$\text{Energy recovery} = 100\,\frac{\Delta H \text{ of } H_2}{\Delta H \text{ of } \mathbf{1a} + \Delta H \text{ of } \mathbf{1b}} \tag{14}$$

## 5. Conclusion and perspective

Sacrificial $H_2$ production can produce $H_2$ in aqueous solutions. Gaseous $H_2$ can be spontaneously isolated from reaction mixture without being separated. Therefore, sacrificial $H_2$ production will provide a promising approach in the utilization of **1a** and **1b** derived from BDF synthesis.

Recent trends are shifting to the development of solar light-responsive photocatalysts. For example, nanotube-type Pt-N/$TiO_2$ (1 wt% Pt) was applied to sacrificial $H_2$ production with **1a** where quantum yield for $H_2$ evolution reached 0.37–0.36 [27]. CuO/$TiO_2$ (1.3 wt% of CuO) was used for sacrificial $H_2$ production using **1a** [40]. Heteroatom (B, N)-doped Pt/$TiO_2$ catalyst produced $H_2$ in 88.7–90.9% yields from **1a** under xenon lamp irradiation [41]. The B, N-doped Pt/$TiO_2$ had absorption in visible light region (400–500 nm). Photo-reforming of **1a** over $CuO_x$/$TiO_2$ (Cu = 0.01–2.8 wt%) gave $H_2$ under visible light irradiation [42]. $H_2$ production was performed over a CuO-$TiO_2$ composite using **1a** and **1b** under sunlight irradiation [43]. Sacrificial $H_2$ production over $Ag_2$/$TiO_2$ from **1a** was performed by irradiation with a xenon lamp [44].

BDF market has significantly increased to adhere to energy and climate policies [45]. If $H_2$ is produced by a photocatalytic process using solar energy and biomass-derived sacrificial agents, it will be the most promising process to construct clean BDF synthesis.

## Conflict of interest

The authors declare that they have no competing interests.

## Author details

Masahide Yasuda[1]*, Tomoko Matsumoto[2] and Toshiaki Yamashita[3]

1 Department of Applied Chemistry, Faculty of Engineering, University of Miyazaki, Miyazaki, Japan

2 Center for Collaborative Research and Community Cooperation, University of Miyazaki, Miyazaki, Japan

3 Department of Chemical Science and Engineering, National Institute of Technology, Miyakonojo College, Miyakonojo, Miyazaki, Japan

*Address all correspondence to: yasuda@cc.miyazaki-u.ac.jp

## References

[1] Navarro M, Peña MA, Fierro JLG. Hydrogen production reactions from carbon feedstocks: Fossil fuels and biomass. Chemical Reviews. 2007;**107**:3952-3991

[2] Yasuda M, Ishii Y, Ohta K. Napiergrass (*Pennisetum purpureum* Schumach) as raw material for bioethanol production: Pretreatment, saccharification, and fermentation. Biotechnology and Bioprocess Engineering. 2014;**19**:943-950

[3] Yasuda M, Takenouchi Y, Nitta Y, Ishii Y, Ohta K. Italian ryegrass (*Lolium multiflorum* lam) as a high potential bio-ethanol resource. Bioenergy Research. 2015;**8**:1303-1309

[4] Ma F, Hanna MA. Biodiesel production: A review. Bioresource Technology. 1999;**70**:1-15

[5] Iulianelli A, Seelam PK, Liguori S, Longo T, Keiski R, Calabr V, et al. Hydrogen production for PEM fuel cell by gas phase reforming of glycerol as byproduct of bio-diesel. The use of a Pd–Ag membrane reactor at middle reaction temperature. International Journal of Hydrogen Energy. 2011;**36**:3827-3834

[6] Fernández Y, Arenillas A, Díez MA, Pis JJ, Menéndez JA. Pyrolysis of glycerol over activated carbons for syngas production J. Journal of Analytical and Applied Pyrolysis. 2009;**84**:145-150

[7] Valliyappan T, Ferdous D, Bakhshi NN, Dalai AK. Production of hydrogen and syngas via steam gasification of glycerol in a fixed-bed reactor. Topics in Catalysis. 2008;**49**:59-67

[8] Wang C, Dou B, Chen H, Song Y, Xu Y, Du X, et al. Hydrogen production from steam reforming of glycerol by Ni–Mg–Al based catalysts in a fixed-bed reactor. Chemical Engineering Journal. 2013;**220**:133-142

[9] Ngo TA, Sim SJ. Dark fermentation of hydrogen from waste glycerol using hyperthermophilic eubacterium *Thermotoga neapolitana*. Environmental Progress & Sustainable Energy. 2012;**31**:466-473

[10] Costa JB, Rossi DM, De Souza EA, Samios D, Bregalda F, Do Carmo M, et al. The optimization of biohydrogen production by bacteria using residual glycerol from biodiesel synthesis. Journal of Environmental Science and Health, Part A. 2011;**46**:1461-1468

[11] Yasuda M, Kurogi R, Tomo T, Shiragami T. Hydrogen production from residual glycerol from biodiesel synthesis by photocatalytic reforming. Journal of the Japan Institute of Energy. 2014;**93**:710-715

[12] Fujishima A, Rao TN, Tryk DA. Titanium dioxide photocatalysis. Journal of Photochemistry and Photobiology C. 2000;**1**:1-21

[13] Galinska A, Walendziewski J. Photocatalytic water splitting over Pt-$TiO_2$ in the presence of sacrificial agents. Energy & Fuels. 2005;**19**:1143-1147

[14] Yasuda M, Matsumoto T, Yamashita T. Sacrificial hydrogen production over $TiO_2$–based photocatalysts: Polyols, carboxylic acids, and saccharides. Renewable and Sustainable Energy Reviews. 2018;**81**:1627-1635

[15] Shiragami T, Tomo T, Matsumoto T, Yasuda M. Structural dependence of alcoholic sacrificial agents on $TiO_2$-photocatalytic hydrogen evolution. Bulletin of the Chemical Society of Japan. 2013;**86**:382-389

[16] Atkins PW. Physical Chemistry. 5th ed. Oxford, UK: Oxford University Press; 1994. pp. 922-926

[17] Shimura K, Yoshida H. Heterogeneous photocatalytic hydrogen production from water and biomass derivatives. Energy & Environmental Science. 2011;**4**:467-481

[18] Yasuda M. Chapter 19: Photocatalytic reforming of lignocelluloses, glycerol, and chlorella to hydrogen. In: Jacob-Lopes E, Zepka LQ, editors. Frontiers in Bioenergy and Biofuels. Rijeka, Croatia: Intech; 2017. pp. 391-406

[19] Yasuda M, Kurogi R, Tsumagari H, Shiragami T, Matsumoto T. New approach to fuelization of herbaceous lignocelluloses through simultaneous saccharification and fermentation followed by photocatalytic reforming. Energies. 2014;**7**:4087-4097

[20] Yasuda M, Takenouchi MY, Kurogi R, Uehara S, Shiragami T. Fuelization of Italian ryegrass and Napier grass through a biological treatment and photocatalytic reforming. Journal of Sustainable Bioenergy Systems. 2015;**5**:1-9

[21] Shiragami T, Tomo T, Tsumagari H, Ishii Y, Yasuda M. Hydrogen evolution from napiergrass by the combination of biological treatment and a Pt-loaded $TiO_2$-photocatalytic reaction. Catalysts. 2012;**2**:56-67

[22] Yasuda M, Hirata S, Matsumoto T. Sacrificial hydrogen production from enzymatic hydrolyzed chlorella over a Pt-loaded $TiO_2$ photocatalyst. Journal of the Japan Institute of Energy. 2014;**95**:599-604

[23] Kennedy JC III, Datye AK. Photochemical heterogeneous oxidation of ethanol over $Pt/TiO_2$. Journal of Catalysis. 1998;**179**:375-389

[24] Salas SE, Rosales BS, de Lasa H. Quantum yield with platinum modified $TiO_2$ photocatalyst for hydrogen production. Applied Catalysis B: Environmental. 2013;**140-141**:523-536

[25] Yasuda M, Kurogi R, Matsumoto T. Quantum yields for sacrificial hydrogen generation from saccharides over a Pt-loaded $TiO_2$ photocatalyst. Research on Chemical Intermediates. 2015;**42**:3919-3928

[26] Gu Q, Fu X, Wang X, Chen S, Leung DY, Xie X. Photocatalytic reforming of C3- polyols for $H_2$ production. Part II. FTIR study on the adsorption and photocatalytic reforming reaction of 2-propanol on $Pt/TiO_2$. Applied Catalysis. B, Environmental. 2011;**106**:689-696

[27] Slamet Tristantini D, Valentina Ibadurrohman M. Photocatalytic hydrogen production from glycerol-water mixture over Pt-N-$TiO_2$ nanotube photocatalyst. International Journal of Energy Research. 2013;**37**:1372-1381

[28] Fu X, Long J, Wang X, Leung Y, Ding Z, Wu L, et al. Photocatalytic reforming of biomass: A systematic study of hydrogen evolution from glucose solution. International Journal of Hydrogen Energy. 2008;**33**:6484-6489

[29] Gomathisankar P, Yamamoto D, Katsumata H, Suzuki T, Kaneco S. Photocatalytic hydrogen production with aid of simultaneous metal deposition using titanium dioxide from aqueous glucose solution. International Journal of Hydrogen Energy. 2013;**38**:5517-5524

[30] Daskalaki VD, Kondarides DI. Efficient production of hydrogen by photo-induced reforming of glycerol at ambient conditions. Catalysis Today. 2009;**144**:75-80

[31] Panagiotopoulou P, Karamerou EE, Kondarides DI. Kinetics and mechanism

of glycerol photo-oxidation and photo-reforming reactions in aqueous $TiO_2$ and $Pt/TiO_2$ suspensions. Catalysis Today. 2013;**209**:46-48

[32] Daskalaki VM, Panagiotopoulou P, Kondarides DI. Production of peroxide species in $Pt/TiO_2$ suspensions under conditions of photocatalytic water splitting and glycerol photoreforming. Chemical Engineering Journal. 2011;**170**:433-439

[33] Slamet R, Gunlazuardi J, Dewi EL. Enhanced photocatalytic activity of Pt deposited on titania nanotube arrays for the hydrogen production with glycerol as a sacrificial agent. International Journal of Hydrogen Energy. 2017;**42**:24014-24025

[34] Bowker M, Davies PR, Al-Mazroai LS. Photocatalytic reforming of glycerol over gold and palladium as an alternative fuel source. Catalysis Letters. 2009;**128**:253-255

[35] Li Y, Lu G, Li S. Photocatalytic hydrogen generation and decomposition of oxalic acid over platinized $TiO_2$. Applied Catalysis, A: General. 2001;**21**:179-185

[36] Li Y, Lu G, Li S. Photocatalytic production of hydrogen in single component and mixture systems of electron donors and monitoring adsorption of donors by in situ infrared spectroscopy. Chemosphere. 2003;**52**:843-850

[37] Mozia S, Heciak A, Morawski AW. The influence of physico-chemical properties of $TiO_2$ on photocatalytic generation of C1-C3 hydrocarbons and hydrogen from aqueous solution of acetic acid. Applied Catalysis. B, Environmental. 2011;**104**:21-29

[38] Zheng X-J, Wei L-F, Zhang Z-H, Jiang Q-J, Wei Y-J, Xie B, et al. Research on photocatalytic $H_2$ production from acetic acid solution by $Pt/TiO_2$ nanoparticles under UV irradiation. International Journal of Hydrogen Energy. 2009;**34**:9033-9041

[39] Yasuda M, Tomo T, Hirata S, Shiragami T, Matsumoto T. Neighboring hetero-atom assistance of sacrificial amines to hydrogen evolution using Pt-loaded $TiO_2$-photocatalyst. Catalysts. 2014;**4**:162-173

[40] Yu J, Hai Y, Jaroniec M. Photocatalytic hydrogen production over CuO-modified titania. Journal of Colloid and Interface Science. 2011;**357**:223-228

[41] Luo N, Jiang Z, Shi H, Cao F, Xiao T, Edwards PP. Photo-catalytic conversion of oxygenated hydrocarbons to hydrogen over heteroatom-doped $TiO_2$ catalysts. International Journal of Hydrogen Energy. 2009;**34**:125-129

[42] Petala A, Ioannidou E, Georgaka A, Bourikas K, Kondarides DI. Hysteresis phenomena and rate fluctuations under conditions of glycerol photo-reforming reaction over $CuOx/TiO_2$ catalysts. Applied Catalysis. B, Environmental. 2015;**178**:201-209

[43] Pai MR, Banerjee AM, Rawool SA, Nayak C, Ehrman SH, Tripathi AK, et al. A comprehensive study on sunlight driven photocatalytic hydrogen generation using low cost nanocrystalline Cu-Ti oxides. Solar Energy Materials & Solar Cells. 2016;**154**:104-120

[44] Wang C, Cai X, Chen Y, Cheng Z, Luo X, Mo S, et al. Efficient hydrogen production from glycerol photoreforming over $Ag_2O$-$TiO_2$ synthesized by a sol-gel method. International Journal of Hydrogen Energy. 2017;**42**:17063-17074

[45] Onwudili JA, Williams PT. Catalytic pyrolysis of low-density polyethylene over alumina-supported noble metal catalysts. Fuel. 2010;**89**:501-509

Section 3

# Use as Additive

Chapter 7

# Inclusion of Crude Glycerin in Diets for Sheep

*Marco Túlio Costa Almeida and Josimari Regina Paschoaloto*

## Abstract

Crude glycerin is the main by-product of biodiesel industry. It has a great potential for reducing the feed costs in ruminant feedlot systems without affecting animal health and performance, mainly as a replacement for conventional food energy sources, such as corn grain. In the past years, great advancements have been achieved with crude glycerin utilization. This by-product is mainly composed of glycerol, an energetic compound of great assimilation by rumen microorganisms, being extensively metabolized in the liver. Recent studies with ovine species have demonstrated that high concentrations of glycerol (more than 76% of crude glycerin) can be used without detrimental effect for animals. In the rumen, glycerol is rapidly metabolized by microorganisms to form volatile fatty acids (VFA), mainly propionate and butyrate. In this way, glycerol constitutes an excellent substrate for gluconeogenesis and animal energy generation. At present, the inclusion of up to 20% of dry matter (DM) in a total diet seems to be the most interesting strategy, as it promotes greatest animal performance. However, other studies suggest that high inclusions of crude glycerin (30% of dry matter) could be possible depending on market price and the structure of farm operation, with favorable economic results.

**Keywords:** by-product, energy source, glycerol, metabolism, performance, small ruminant

## 1. Introduction

An increasing worldwide interest in alternative fuel sources and in a more diversified energy matrix has led researchers to search for alternatives to fossil fuels [1, 2]. In this scenario, biodiesel appears as the main substitute for this energetic matrix. In addition to being a renewable energy source, biodiesel has also lower power pollution [3]. However, because of government incentives, there was a huge growth in the production of this biofuel. Consequently, the availability of by-products (cakes, meals, and glycerin) also increased, being now considered a residual with disposal costs [4]. With this concern, the use of crude glycerin in animal diets has been explored as an economically and environmentally acceptable way to utilize this by-product.

Crude glycerin represents the main by-product of biodiesel production. Approximately 10% of the total volume of biodiesel produced becomes crude glycerin [5, 6]. The energy content of crude glycerin turns it into a promising ingredient for animal feed. The energy value of crude glycerin is similar to corn grain, being estimated as 3.47 Mcal/kg of dry matter (DM) for ruminants [7]. In this way, this by-product has been studied as a macro-ingredient since the 1950s [8, 9].



IntechOpen

Crude glycerin is considered safe for use as animal feed since 2016 [10]. Due to the increased availability and relatively low cost, it became again the focus of studies as an alternative energy source in ruminants' diets [1, 11, 12], with promising results.

Crude glycerin is mainly composed of glycerol (more than 76%), an energetic compound of great assimilation by rumen microorganisms, being extensively metabolizable in the liver [13]. Once ingested and in the animal rumen, glycerol mostly disappears in the first 24 h [14]. It may be directly absorbed by the epithelium of the digestive system via passive diffusion (probably without facilitated diffusion carriers) and act as a gluconeogenic substrate in the liver [15, 16]. Glycerol may be transformed into propionate and butyrate [17, 18] by ruminal microorganisms, or it may outflow from the rumen through the omasal orifice without transformation [15, 16].

The fermentation of crude glycerin may promote better stability to rumen environment preventing metabolic disorders. It could also avoid severe reduction of ruminal pH and the development of rumen acidosis. These preventive actions are possible mainly by the reduction in lactic acidosis due to an increase in the population of lactate-consuming bacteria and undue fermentations [19].

On the other hand, crude glycerin may have a detrimental effect on the growth of structural carbohydrate-fermenting bacteria [17, 20]. Previous studies have shown a negative impact of crude glycerin on cellulolytic bacteria growth of [20], resulting in reductions in fiber digestibility [12, 21, 22], consequently decreasing DM intake and methane production [23]. However, in other studies these effects on rumen fermentation were not observed, as reported by [24, 25] with inclusions up to 12% of crude glycerin.

There are many controversies and inconsistent outcomes about crude glycerin inclusion in ruminant diets. Most of them are related to the end products of glycerol fermentation and its effects on ruminal microorganisms. However, promising results regarding growth performance and feed efficiency have been reported for inclusion up to 20% of crude glycerin in diets for feedlot lambs [26–28]. Therefore, this chapter discusses important aspects of different inclusions and associations of crude glycerin, including glycerol digestibility and metabolism and practical applications in sheep nutrition and production.

## 2. Crude glycerin

In general, glycerin is derived from renewable biological sources (such as vegetable oils or animal fats), occurring as triglycerides. Hydrolysis of these triglycerides results in crude glycerin and fatty acid (**Figure 1**). However, crude glycerin from animal source is prohibited for ruminants. A ban feeding ruminants with animal residues to ruminants is due to mad cow disease (bovine spongiform

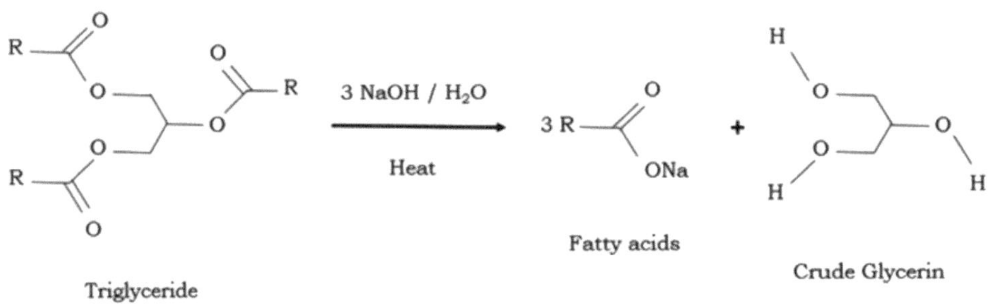


**Figure 1.**
*Hydrolysis of triglyceride to fatty acids and crude glycerin.*

| Reference | Glycerol | Protein | Fat | Ash | Salts | Na | Methanol |
|---|---|---|---|---|---|---|---|
| Mach et al. [7] | 857.0 | — | — | — | 55.0 | — | 0.09 |
| Egea et al. [60] | 875.0 | — | — | 59.0 | — | 20.0 | 0,05 |
| Eiras et al. [64] | 812.0 | — | — | 47.5 | — | — | 0.03 |
| Paschoaloto et al. [74] | 860.0 | 12.0 | 2.00 | — | 60.0 | — | <0.01 |
| Almeida et al. [26] | 830.0 | 11.0 | — | — | 60.0 | — | <0.01 |

**Table 1.**
*Composition and nutritional value of crude glycerin used in ruminant nutrition (g/kg DM).*

encephalopathy). This neurodegenerative disease has been linked to the practice of feeding the cattle with meat-and-bone meal putatively contaminated with scrapie agent [29]. For this reason, only crude glycerin derived from vegetable oils can be used for ruminant feeding.

Approximately 10% of the total volume of biodiesel produced becomes crude glycerin [5, 6]. Crude glycerin is a colorless, hygroscopic, and sweet-tasting viscous liquid. It is composed of glycerol and trace amounts of water and methanol. However, the chemical composition is dependent on glycerin purity and on the oil source utilized [30]. On average, crude glycerin contains 83% glycerol, 0.4% fat, 0.9% crude protein, 5.3% ash, 7.3% salts, and <0.05% of methanol (**Table 1**).

The energy value of crude glycerin is similar to corn grain, being estimated as 3.47 Mcal/kg DM for ruminants [7]. For this reason, crude glycerin becomes a promising ingredient for ruminant feed.

Methanol content in crude glycerin is very low, and in a total mixed ration, it is even lower. Thus, there are no problems associated with ingestion of methanol from crude glycerin by ruminants. According to [31], the high health risk associated with methanol consumption due to the inclusion of crude glycerin in diets is not expected in ruminant because methanol can be naturally produced in the rumen as a result of pectin fermentation. Normally, due to the type of crude glycerin used in ruminant diets (with more than 76% glycerol), it is not a problem. In Ref. [32], authors observed problems in the performance of feedlot lambs related to high impurity contents in crude glycerin (i.e., methanol, NaCl), even at low inclusions (12% DM). However, besides the high impurity, the crude glycerin used had low content of glycerol (<40%), which is not common to be seen on the market. Anyway, the methanol content should be monitored to avoid future problems.

## 3. Ruminal digestion and fermentation of crude glycerin

Glycerol constitutes an excellent substrate for gluconeogenesis and animal energy generation. It can be converted to glucose in the liver providing energy for cellular metabolism [33]. Once ingested, glycerol is totally metabolizable by the animal in the first 24 h [14]. Other authors reported even faster metabolization by microorganisms between 4 and 6 h [34]. It demonstrates the potential of crude glycerin as a rapid energetic source for ruminants.

Into the rumen, glycerol may be fermented and transformed in volatile fatty acids (VFA) by ruminal microorganisms (25–45%) or be directly absorbed through the rumen papillae (~43%) or escape the rumen by outflow through the omasal orifice without transformation due to its high solubility [15].

Glycerin fermentation changes the VFA profile of the rumen (total concentration and molar proportion) because it is preferentially converted to propionate and butyrate [35, 36]. Thus, the acetate/propionate rate is altered [37–39]. The proportion of propionate and butyrate generally increases at the expense of acetate when diets are supplemented with crude glycerin. Recent studies with ruminal fermentation continuous culture [25] and feedlot goats [40] showed linear increases in propionic acid with linear decreases in acetic acid concentrations when crude glycerin replaced dietary corn up to 20%. No changes were observed for total VFA concentration in this work. However, several in vivo and in situ studies have reported increases in total VFA production [34, 41–43].

Differently, other authors evaluating the increasing inclusions of crude glycerin in diets for crossbreed lambs (up to 30% DM, in [28, 44]) and Nellore bulls (up to 30% DM, [36]) observed a linear decrease in total VFA with no changes in propionic acid concentration. However, a pronounced reduction in acetic acid concentration was observed. These results can be explained mainly due to the deleterious effect of crude glycerin on rumen fibrolytic microorganisms, as described by [13, 20]. They reported reductions on *Butyrivibrio fibrisolvens* and *Ruminococcus flavefaciens* bacterial groups. These authors have also found a decrease in neutral detergent fiber (NDF) digestibility and in *Selenomonas ruminantium* and *Clostridium protoclasticum* population by increasing concentrations of glycerin. However, no effects were observed in rumen pH, in $NH_3$-N concentrations, and in DM digestibility.

Reduction in methane ($CH_4$) production has been reported as a detrimental effect of crude glycerin inclusion on the growth of structural carbohydrate-fermenting bacteria. It also happened due to low production of total VFA and acetic acid in the rumen [28, 36, 45]. This reduction probably occurs because the amount of $H_2$ in the rumen decreases. Hence, availability for methanogenic microorganisms is used to reduce $CO_2$ to $CH_4$ [46, 47]. For this reason, the inclusion of crude glycerin in ruminants' diets becomes an interesting alternative for mitigating $CH_4$ emission by livestock.

The deleterious effect of crude glycerin on fiber fermentation was confirmed by [28, 36] when high inclusions of crude glycerin (up to 30% DM) in diets for feedlot crossbred lambs and Nellore bulls were studied, respectively. The authors observed a linear decrease in fiber fraction in vitro total tract digestibility when crude glycerin was included. Although the absolute yield of propionate was not affected by the increasing inclusion of crude glycerin, a linear increase in propionate proportion was observed.

The change in propionate relative proportion happened, probably, due to crude glycerin fermentation characteristics. Glycerin is fermented, mainly by bacteria of the genus *Selenomonas*, being mostly used by animals in the first hours after ingestion [7, 48]. Propionate produced in the rumen is readily absorbed through the ruminal wall, getting into the portal vein. Then, when reaching the liver, it is transformed into glucose and is called glycogenic VFA. The glucogenic pathway in ruminants is shown in **Figure 2**.

The mechanisms that determine the VFA production in quantity and quality can influence the rumen pH and their buffering [49]. When energetic ingredients (starch or soluble sugars) are added to ruminant diets, a reduction in ruminal pH is frequently observed. When ruminal pH is lower than 5.6, strong organic acids (like lactate, pKa = 3.86) are produced on a major scale, which can cause a nutritional disorder, named ruminal acidosis. The consequences of ruminal acidification process can range from decrease of DM intake up to the animal's death.

Crude glycerin can prevent metabolic disorders, avoiding severe reduction on ruminal pH and the development of rumen acidosis due to glycerin fermentation profile. Inclusions in ruminal pH were observed when lambs were fed with up to 20 or 30% of crude glycerin [28, 44], respectively. During glycerol ruminal fermentation, there is no generation of lactic acid. This happens because propionic acid is produced by an alternative fermentative pathway via succinate [50]. For this reason,

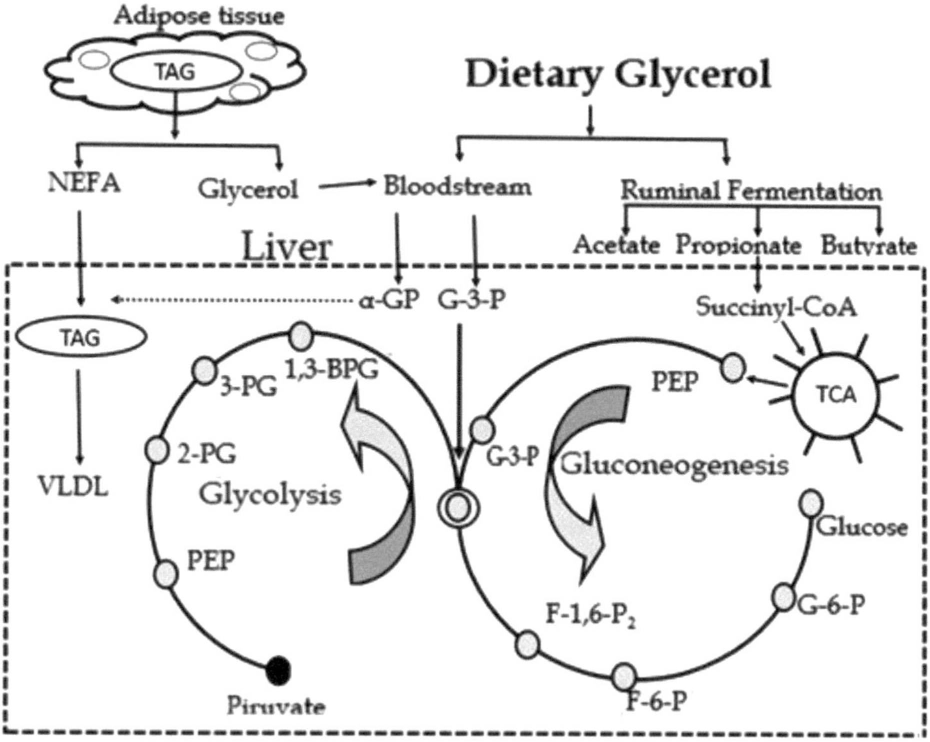


**Figure 2.**
*Proposed metabolism of glycerol in ruminant animals (adapted of Osman et al. [41]).*

crude glycerin may promote better stability to rumen environment compared with starch-rich ingredients. Lactic acidosis is reduced due to an increase in the population of lactate-consuming bacteria and undue fermentation [19]. As a consequence, some benefits are gained, such as rumen development and feed efficiency improvement [15, 16].

Authors in Ref. [50] evaluated the effects of increasing inclusions of crude glycerin (up to 30% DM) in diets for crossbred lambs in two different periods of feedlot (adaptation and finishing). The authors observed that crude glycerin did not compromise the stomach compartments and rumen papillae measurements in both periods of feedlot. No clinical manifestations resulted from ruminal acidosis (such as liver abscess, ruminitis, and lesions in ruminal mucosa) in any period and treatment studied. The authors concluded that the replacement of corn grain by crude glycerin (up to 30% DM) was effective in the animals' adaptation to concentrate-based diets. The good results were possible due to the better ruminal conditions provided by crude glycerin fermentation.

## 4. Effects on dry matter intake and growth performance

The use of crude glycerin has been investigated in diets for feedlot lambs [1, 27, 51–53]. The results are varied, regarding intake, digestibility, and performance. Probably, this may be related to the degree of purity of the crude glycerin used.

Authors in Ref. [1] evaluated the effects of high concentrations of crude glycerin (up to 30% DM basis) in diets for feedlot lambs. The authors reported positive results in animal performance when crude glycerin was included in 10% DM. Linear reductions in dry matter intake (DMI) were also observed by the authors. However, when all crude glycerin treatments were compared with the control diet (10, 20, and 30% vs. control diet), no significant difference was observed. Authors [42, 54, 55], working with feedlot beef cattle, also observed a reduction in DMI when levels of glycerin were higher than 10%. In others works, even in low concentrations (<10% DM), the inclusion of crude glycerin promoted a drop in DMI for lambs and goats [32, 56, 57].

The lower value of DMI observed when crude glycerin was included in the diets may be explained by the deleterious selection of fibrolytic microorganisms, known to be sensitive to glycerin. Previous studies have shown that glycerol provides a selection of rumen microorganisms, mainly fibrolytic ones, affecting NDF digestibility. The reduction in NDF digestibility is attributed to the decrease in DNA concentration from bacteria *Selenomonas ruminantium* and *Butyrivibrio fibrisolvens* [13]. Furthermore, according to [20], glycerin decreases microorganism growth, cell membrane permeability, and adhesion of bacteria in feed. Thus, these facts can reduce the feed passage rate through the rumen. This results in ruminal filling, known as repletion state [58], consequently reducing DMI. Another plausible explanation for DMI reduction is the fact that ingestion of glycerol can improve animal metabolic status. This improvement occurs due to the increased energy intake from glucose with increased propionate and reduced acetate/propionate ratio in the rumen, satiating the animal in terms of energy [1, 15].

Authors in Ref. [40] fed goats with up to 20% DM and in Ref. [51] fed lambs with up to 30% DM of crude glycerin. The authors, contradicting the above mentioned, observed no difference in DMI in any inclusions offered to the animals. On the other hand, authors in Ref. [27] reported that inclusion of crude glycerin up to 15% DM in diets of finishing lambs increased DMI and feedlot performance in the first 14 days. According to these authors, the controversies observed in DMI are probably due to the purity level of the glycerin used. Other important factors were the forage/concentrate ratio, the amount of dietary starch and fiber used, and the tolerance and acceptability of the ingredient by each animal species. Thus, the best crude glycerin association should be used to avoid deleterious effects.

Nevertheless, despite the reductions in DMI, authors in Ref. [59] reported average daily gain (ADG) of 250 g/day with the inclusion of 12.5% DM of crude glycerin. Similarly, authors in Ref. [1] observed ADG of 332 g/day with the inclusion of 10% DM. In both studies, the crude glycerin used contained ~83% of glycerol, and the forage/concentrate ratio used was 40:60, the corn silage being the forage source. Furthermore, other researchers reported improvements in feed efficiency when crude glycerin was included up to 12% DM in diets for sheep and beef cattle [54, 55, 57].

## 5. Effects on carcass characteristics and meat quality

Recent studies did not present negative effects of crude glycerin inclusion on carcass and meat traits [7, 27, 60]. The inclusion up to 30% crude glycerin in diets for feedlot lambs promoted lower carcass weights [1, 61]. However, other studies have demonstrated that the inclusion of this by-product has effects on carcass grade [1, 62]; increases the intramuscular fat and oleic acid content [63]; decreases myristic, palmitic, and stearic acids in *Longissimus muscle* [64]; increases the monounsaturated fatty acid (MUFA) content and conjugated linoleic acid content [65]; decreases saturated fatty acid; and increases unsaturated and odd-chain fatty acid contents [62].

Carcass and meat traits can be altered in the meat of animals fed with crude glycerin by increasing marbling deposition as a function of glycogenic precursor absorption [66]. The change in these features can also be achieved by increasing unsaturated fatty acid content on meat [63, 64, 67] possibly due to ruminal lipolysis inhibition [19, 68].

It is known that inclusion of crude glycerin in diets decreased the DNA concentration for *Selenomonas ruminantium* and *Butyrivibrio proteoclasticus* [13, 17], important bacteria in biohydrogenation process [69–72]. This fact suggests that when glycerol is incorporated in diets, it may reduce lipolysis and biohydrogenation in the rumen, promoting a greater flow of unsaturated fatty acids to be incorporated in meat [19, 68]. For these reasons, the crude glycerin inclusion in diets for ruminants promotes a healthier meat for humans.

## 6. Crude glycerin as a strategy ingredient

Crude glycerin has been recently studied as a great ingredient to prevent metabolic disorders when high inclusions of concentrate were offered to animals in feedlot systems [26, 50, 73, 74]. Because of its own fermentation profile, crude glycerin can avoid severe reduction of ruminal pH and the development of rumen acidosis. The larger portion of crude glycerin (~43%) is rapidly absorbed by rumen papillae, whereas 25–45% are fermented to butyrate and propionate by alternative fermentative pathway (via succinate) and do not generate lactic acid, benefiting the rumen development and thus improving the feed efficiency [15, 16, 50].

When animals are submitted to a feedlot management, abrupt changes occur in their diets. Generally, high inclusions of starch-rich cereal grains are used. Then, a correct adaptation to diets should be done, mainly in the first 2 weeks [75]. When animals are not correctly adapted to diets, changes in feeding behavior with reduced DMI [76]; clinical manifestations such as laminitis, liver abscesses, and ruminitis; and inappropriate ruminal fermentation [77, 78] are prevalent. Recent studies have shown that crude glycerin is effective in the animals' adaptation to concentrate-based diets [27, 50]. Authors in Ref. [50] evaluated the effects of increasing inclusions of crude glycerin (up to 30% DM) in diets for crossbred feedlot lambs in two different periods (adaptation and finishing). The authors observed that crude glycerin did not compromise the stomach compartments and rumen papillae measurements in both periods of feedlot. No clinical manifestations resulted from ruminal acidosis in any period and treatment studied. Authors in Ref. [27] reported that the feed efficiency during the first 14 days of the feedlot was greater for all glycerin-fed lambs when compared with lambs not fed with glycerin.

Apparently, crude glycerin may have stimulated rumen growth due to its fermentative characteristic, mainly by increasing the total rumen VFA [14]. The presence of VFA in ruminal lumen promotes the growth and proliferation of papillae. With increased absorption area, the capacity of removing these acids is improved as well as the provision of greater energy absorption for the animal [79–81], thus improving the animal performance. In this way, crude glycerin presents good aspects to the adaptation physiological process and can be indicated to adapt the animal to concentrate-based diets, mainly in the first 2 weeks of the feedlot.

## 7. Crude glycerin practical recommendations

Because of its high viscosity (~8.5 cSt), crude glycerin can be used to change the physical form of diets. It helps in particle aggregation, thus facilitating the feed intake. Inclusions up to 30% can be used for this purpose. However, depending on the diet ingredients composition, an accumulation of the ingredient in the feed bunk bottom may occur.

## 8. Conclusions

Based on what has been discussed in this chapter, crude glycerin can be considered as a good energy ingredient for sheep nutrition. At present, inclusions up to 20% DM in a total diet seem to be the most interesting strategy. They promote greater animal performance and a healthier meat for humans. However, other studies suggest that higher inclusions of crude glycerin (30% of dry matter) could be possible depending on market price and the farm operation structure with favorable economic results.

## Conflict of interest

The authors declare no known conflict of interest.


## Author details

Marco Túlio Costa Almeida* and Josimari Regina Paschoaloto
Department of Animal Science, Federal University of Piauí (UFPI), Bom Jesus, Piauí, Brazil

*Address all correspondence to: marco.almeida@ufpi.edu.br

## References

[1] Almeida MTC, Ezequiel JMB, Paschoaloto JR, Perez HL, Carvalho VB, Castro Filho ES, et al. Effects of high concentrations of crude glycerin in diets for feedlot lambs: Feeding behaviour, growth performance, carcass and non-carcass traits. Animal Production Science. 2017;**58**:1271-1278. DOI: 10.1071/AN16628

[2] van Cleef EHCB, D'Aurea AP, Favaro VR, van Cleef FOS, Barducci RS, Almeida MTC, et al. Effects of dietary inclusion of high concentrations of crude glycerin on meat quality and fatty acid profile of feedlot fed Nellore bulls. PLoS One. 2017;**12**:e0179830. DOI: 10.1371/journal.pone.0179830

[3] Lima DO, Sogabe VP, Calarge TCC. Uma Análise sobre o Mercado Mundial do biodiesel. CPMark UNIMEP. 2014;**2**(1):44-59

[4] Yazdani SS, Gonzalez R. Anaerobic fermentation of glycerol: A path to economic viability for the biofuel industry. Current Opinion in Biotechnology. 2007;**18**:213-219. DOI: 10.1016/j.copbio.2007.05.002

[5] Dasari MA, Kiatsimkul PP, Sutterlin WR, Suppes GJ. Low-pressure hydrogenolysis of glycerol to propylene glycol. Applied Catalysis A: General. 2005;**281**:225-231. DOI: 10.1016/j.apcata.2004.11.033

[6] Johnson DT, Taconi KA. The glycerin glut: Options for the value-added conversion of crude glycerol resulting from biodiesel production. Environmental Progress & Sustainable Energy. 2007;**26**:338-348. DOI: 10.1002/ep.10225

[7] Mach N, Bach A, Devant M. Effects of crude glycerin supplementation on performance and meat quality of Holstein bulls fed high-concentrate diets. Journal of Animal Science. 2009;**87**:632-638. DOI: 10.2527/jas.2008-0987

[8] Garton GA, Lough AK, Vioque E. Glyceride hydrolysis and glycerol fermentation by sheep rumen contents. Journal of General Microbiology. 1961;**25**:215-225. DOI: 10.1099/00221287-25-2-215

[9] Johns A. Fermentation of glycerol in the ruminal of sheep. New Zealand Journal of Science and Technology. 1953;**35**:262-269

[10] FDA. Food Drug Administration code of Federal Regulations, 21CFR582.1320. Good Laboratory Practice. 2006;**21**:306-319

[11] Ezequiel JMB, Sancanari JBD, Neto ORM, da Silva ZF, Almeida MTC, Silva DAV, et al. Effects of high concentrations of dietary crude glycerin on dairy cow productivity and milk quality. Journal of Dairy Science. 2015;**98**:8009-8017. DOI: 10.3168/jds.2015-9448

[12] van Cleef EHCB, Ezequiel JMB, D'aurea AP, Fávaro VR, Sancanari JBD. Crude glycerin in diets for feedlot Nellore cattle. Revista Brasileira de Zootecnia. 2014;**43**:86-91. DOI: 10.1590/S1516-35982014000200006

[13] Abo El-Nor S, AbuGhazaleh AA, Potu RB, Hastings D, Khattab MSA. Effects of differing levels of glycerol on rumen fermentation and bacteria. Animal Feed Science and Technology. 2010;**162**:99-105. DOI: 10.1016/j.anifeedsci.2010.09.012

[14] Trabue S, Scoggin K, Tjandrakusuma S, Rasmussen MA, Reilly PJ. Ruminal fermentation of propylene glycol and glycerol. Journal of Agricultural and Food Chemistry. 2007;**55**:7043-7051. DOI: 10.1021/jf071076i

[15] Krehbiel CR. Ruminal and physiological metabolism of glycerin. Journal of Animal Science. 2008;**86**(E-Suppl. 2):392

[16] Omazic AW, Kronqvist C, Zhongyan L, Martens H, Holtenius K. The fate of glycerol entering the rumen of dairy cows and sheep. Journal of Animal Physiology and Animal Nutrition. 2015;**99**:258-264. DOI: 10.1111/jpn.12245

[17] AbuGhazaleh AA, Abo El-Nor S, Ibrahim SA. The effect of replacing corn with glycerol on ruminal bacteria in continuous culture fermenters. Journal of Animal Physiology and Animal Nutrition. 2011;**95**(3):313-319

[18] Donkin SS. Glycerol from biodiesel production: The new corn for dairy cattle. Revista Brasileira de Zootecnia. 2008;**37**(supl. especial):280-286. DOI: 10.1590/S1516-35982008001300032

[19] Krueger NA, Anderson RC, Tedeschi LO, Callaway TR, Edrington TS, Nisbet DJ. Evaluation of feeding glycerol on free-fatty acid production and fermentation kinetics of mixed ruminal microbes in vitro. Bioresource Technology. 2010;**101**:8469-8472. DOI: 10.1016/j.biortech.2010.06.010

[20] Roger V, Fonty G, Andre C, Gouet P. Effects of glycerol on the growth, adhesion, and cellulolytic activity of rumen cellulolytic bacteria and anaerobic fungi. Current Microbiology. 1992;**25**:197-201. DOI: 10.1007/BF01570719

[21] Donkin SS, Koser SL, White HM, Doane PH, Cecava MJ. Feeding value of glycerol as a replacement for corn grain in rations fed to lactating dairy cows. Journal of Dairy Science. 2009;**92**(10):5111-5119

[22] Shin JH, Wang D, Kim SC, Adesogan AT, Staples CR. Effects of feeding crude glycerin on performance and ruminal kinetics of lactating Holstein cows fed corn silage or cottonseed hull-based, low-fiber diets. Journal of Dairy Science. 2012;**95**(7):4006-4016. DOI: 10.3168/jds.2011-5121

[23] Holter JB, Young AJ. Methane prediction in dry and lactating Holstein cows. Journal of Dairy Science. 1992;**75**(8):2165-2175

[24] Beck R. Effects of Dietary Glycerol Supplementation on Growth, Carcass Traits, Nutrient Digestibility, and Rumen Function in Finishing Lambs. Ann Arbor, MI, USA: South Dakota State University; 2011

[25] Ramos MH, Kerley MS. Effect of dietary crude glycerol level on ruminal fermentation in continuous culture and growth performance of beef calves. Journal of Animal Science. 2012;**90**:892-899. DOI: 10.2527/jas.2011-4099

[26] Almeida MTC, Ezequiel JMB, Paschoaloto JR, Carvalho VB, Perez HL, Fávaro VR, et al. Crude glycerin combined with food additives in feeding beef cattle. Revista Brasileira de Zootecnia. 2018;**47**:e20170124. DOI: 10.1590/rbz4720170124

[27] Gunn PJ, Schultz AF, Van Emon ML, Neary MK, Lemenager RP, Rusk CP, et al. Effects of elevated crude glycerin concentrations on feedlot performance, carcass characteristics, and serum metabolite and hormone concentrations in finishing ewe and wether lambs. The Professional Animal Scientist. 2010;**26**:298-306. DOI: 10.15232/S1080-7446(15)30597-0

[28] van Cleef EHCB, Almeida MTC, Perez HL, Paschoaloto JR, Castro Filho ES, Ezequiel JMB. Effects of partial or total replacement of corn cracked grain with high concentrations of crude glycerin on rumen metabolism of crossbred sheep. Small Ruminant Research. 2018;**159**:45-51. DOI: 10.1016/j.smallrumres.2017.12.011

[29] Taylor DM, Woodgate SL, Fleetwood AJ, Cawthorne RJG. Effect of rendering procedures on the scrapie agent. The Veterinary Record. 1997;**141**(25):643-649

[30] Thompson JC, He B. Characterization of crude glycerol from biodiesel production from multiple feedstocks. Applied Engineering in Agriculture. 2006;**22**(2):261-265. DOI: 10.13031/2013.20272

[31] Pol A, Demeyer DI. Fermentation of methanol in the sheep rumen. Applied and Environmental Microbiology. 1988;**54**(3):832-834

[32] Lage JF, Paulino PVR, Pereira LGR, Valadares Filho SC, Oliveira AS, Detmann E, et al. Glicerina bruta na dieta de cordeiros terminados em confinamento. Pesquisa Agropecuária Brasileira. 2010;**45**:1012-1020. DOI: 10.1590/S0100-204X2010000900011

[33] Goff JP, Horst RL. Role of acid-base physiology on the pathogenesis of parturient hypocalcaemia (milk fever)—The DCAD theory in principal and practice. Acta Veterinaria Scandinavica. Supplementum. 2003;**97**:51-56

[34] Remond B, Souday E, Jouany JP. In vitro and in vivo fermentation of glycerol by rumen microbes. Animal Feed Science and Technology. 1993;**41**:121-132. DOI: 10.1016/0377-8401(93)90118-4

[35] San Vito E, Lage J, Messana J, Dallantonia E, Frighetto R, Reis R, et al. Performance and methane emissions of grazing Nellore bulls supplemented with crude glycerin. Journal of Animal Science. 2016;**94**:4728-4737. DOI: 10.2527/jas.2016-0530

[36] van Cleef EHCB, Almeida MTC, Perez HL, van Cleef FOS, Silva DAV, Ezequiel JMB. Crude glycerin changes ruminal parameters, in vitro greenhouse gas profile, and bacterial fractions of beef cattle. Livestock Science. 2015;**178**:158-164. DOI: 10.1016/j.livsci.2015.06.016

[37] Carvalho ER, Schmelz-Roberts NS, White HM, Doane PH, Donkin SS. Replacing corn with glycerol in diets for transition dairy cows. Journal of Dairy Science. 2011;**94**(2):908-916

[38] Kristensen NB, Raun BML. Ruminal and intermediary metabolism of propylene glycol in lactating Holstein cows. Journal of Dairy Science. 2007;**90**(10):4707-4717

[39] Linke PL, DeFrain JM, Hippen AR, Jardon PW. Ruminal and plasma responses in dairy cows to drenching or feeding glycerol. Journal of Dairy Science. 2004;**87**:49-60

[40] Chanjula P, Pakdeechanuan P, Wattanasit S. Effects of dietary crude glycerin supplementation on nutrient digestibility, ruminal fermentation, blood metabolites, and nitrogen balance of goats. Asian-Australasian Journal of Animal Sciences. 2014;**27**(3):365

[41] Osman MA, Allen PS, Mehyar NA, Bobe G, Coetzee JF, Koehler KJ, et al. Acute metabolic responses of postpartal dairy cows to subcutaneous glucagon injections, oral glycerol, or both. Journal of Dairy Science. 2008;**91**(9):3311-3322

[42] Schröder A, Südekum KH. Glycerol as a by-product of biodiesel production in diets for ruminants. In: Wratten N, Salisbury PA, editors. New Horizons for an Old Crop Proceedings 10th International Rapeseed Congress; New South Wales, Australia. Paper. No. 241. Gosford: The Regional Institute Ltd.; 1999

[43] Wang C, Liu Q, Huo WJ, Yang WZ, Dong KH, Huang YX, et al. Effects of glycerol on rumen fermentation, urinary excretion of purine derivatives and feed digestibility in steers. Livestock

Science. 2009;**121**:15-20. DOI: 10.1016/j.livsci.2008.05.010

[44] Polizel DM, Gentil RS, Ferreira EM, Souza RA, Freire APA, Faleiro Neto JA, et al. Rumen metabolism in lambs fed high-concentrate diets containing increasing levels of crude glycerin. Journal of Animal Science. 2013;**91**(Suppl. 2):96

[45] Homem Junior AC, Ezequiel JMB, Fávaro VR, Almeida MTC, Paschoaloto JR, D'Áurea AP, et al. Methane production by in vitro ruminal fermentation of feed ingredients. Semina: Ciências Agrárias. 2017;**38**(2):877-884

[46] Janssen PH. Influence of hydrogen on rumen methane formation and fermentation balances through microbial growth kinetics and fermentation thermodynamics. Animal Feed Science and Technology. 2010;**160**(1-2):1-22

[47] Van Soest PJ. Nutritional Ecology of the Ruminant. 2nd ed. Ithaca, New York: Cornell University Press; 1994. 488 p

[48] Ferraro SM, Mendoza GD, Miranda LA, Gutierrez CG. In vitro gas production and ruminal fermentation of glycerol, propylene glycol and molasses. Animal Feed Science and Technology. 2009;**154**(1-2):112-118. DOI: 10.1016/j.anifeedsci.2009.07.009

[49] Millen DD, Pacheco RDL, Cabral LC, Cursino LL, Watanabe DHM, Rigueiro ALN. Ruminal acidosis. In: Millen DD, Arrigoni MDB, Pacheco RDL, editors. Rumenology. Switzerland: Springer Nature; 2016. pp. 127-156

[50] Almeida MT, Ezequiel JM, Paschoaloto JR, Perez HL, Carvalho VB, Castro Filho ES, et al. Rumen and liver measurements of lambs fed with high inclusions of crude glycerin in adaptation and finishing period of feedlot. Small Ruminant Research. 2018;**167**:1-5

[51] Gomes MAB, Moraes GV, Mataveli M, Macedo FAF, Carneiro TC, Rossi RM. Performance and carcass characteristics of lambs fed on diets supplemented with glycerin from biodiesel production. Brazilian Journal of Animal Science. 2011;**40**(10):2211-2219. DOI: 10.1590/S1516-35982011001000022

[52] Lage JF, Paulino PVR, Pereira LGR, Duarte MS, Valadares Filho SC, Oliveira AS. Carcass characteristics of feedlot lambs fed crude glycerin contaminated with high concentrations of crude fat. Meat Science. 2014;**96**(1):108-113. DOI: 10.1016/j.meatsci.2013.06.020

[53] Musselman AF, Van Emon ML, Gunn PJ, et al. Effects of crude glycerin on feedlot performance and carcass characteristics of market lambs. In: Proceedings, Western Section, American Society of Animal Sciences; Purdue University Department of Youth Development and Agricultural Education and Department of Animal Sciences; 2008. pp. 353-355

[54] Parsons GL, Shelor MK, Drouillard JS. Performance and carcass traits of finishing heifers fed crude glycerin. Journal of Animal Science. 2009;**87**:653-657. DOI: 10.2527/jas.2008-1053

[55] Pyatt NA, Doane PH, Cecava MJ. Effect of crude glycerin in finishing cattle diets. Journal of Animal Science. 2007;**85**(1):530

[56] Chanjula P, Raungprim T, Yimmongkol S, Poonko S, Majarune S, Maitreejet W. Effects of elevated crude glycerin concentrations on feedlot performance and carcass characteristics in finishing steers. Asian-Australasian Journal of Animal Sciences. 2016;**29**(1):80

[57] Andrade GP, de Carvalho FFR, Batista ÂMV, Pessoa RAS, da Costa CA, Cardoso DB, et al. Evaluation of crude glycerin as a partial substitute of corn

grain in growing diets for lambs. Small Ruminant Research. 2018;**165**:41-47

[58] Allen MS. Effects of diet on short-term regulation of feed intake by lactating dairy cattle. Journal of Dairy Science. 2000;**83**:1598-1624. DOI: 10.3168/jds.S0022-0302(00)75030-2

[59] Romanzini EP, Sobrinho AGS, Valença RL, Borghi TH, de Andrade N, Bernardes PA. Feedlot of lambs fed biodiesel co-products: Performance, commercial cuts and economic evaluation. Tropical Animal Health and Production. 2018;**50**(1):155-160

[60] Egea M, Linares MB, Garrido MD, Villodre C, Madrid J, Orengo J, et al. Crude glycerine inclusion in Limousin bull diets: Animal performance, carcass characteristics and meat quality. Meat Science. 2014;**98**(4):673-678

[61] Borghi TH, Silva Sobrinho AGD, Zeola NMBL, Almeida FAD, Cirne LGA, Lima ARC. Dietary glycerin does not affect meat quality of Ille de France lambs. Revista Brasileira de Zootecnia. 2016;**45**(9):554-562

[62] Carvalho VB, Leite RF, Almeida MT, Paschoaloto JR, Carvalho EB, Lanna DP, et al. Carcass characteristics and meat quality of lambs fed high concentrations of crude glycerin in low-starch diets. Meat Science. 2015;**110**:285-292. DOI: 10.1016/j.meatsci.2015.08.001

[63] Carvalho JRR, Chizzotti ML, Ramos EM, Neto OM, Lanna DPD, Lopes LS, et al. Qualitative characteristics of meat from young bulls fed different levels of crude glycerin. Meat Science. 2014;**96**(2):977-983

[64] Eiras CE, Marques JA, Prado RM, Valero MV, Bonafé EG, Zawadzki F. Glycerine levels in the diets of crossbred bulls finished in feedlot: Carcass characteristics and meat quality. Meat Science. 2014;**96**(2):930-936. DOI: 10.1016/j.meatsci.2013.10.002

[65] Lage JF, Berchielli TT, San Vito E, Silva RA, Ribeiro AF, Reis RA. Fatty acid profile, carcass and meat quality traits of young Nellore bulls fed crude glycerin replacing energy sources in the concentrate. Meat Science. 2014;**96**(3):1158-1164. DOI: 10.1016/j.meatsci.2013.10.027

[66] Schoonmaker JP, Cecava MJ, Faulkner DB, Fluharty FL, Zerby HN, Loerch SC. Effect of source of energy and rate of growth on performance, carcass characteristics, ruminal fermentation, and serum glucose and insulin of early-weaned steers. Journal of Animal Science. 2003;**81**(4):843-855

[67] Favaro VR, Ezequiel JMB, Almeida MTC, D'Aurea AP, Paschoaloto JR, van Cleef EHCB, et al. Carcass traits and meat quality of Nellore cattle fed different non-fiber carbohydrates sources associated with crude glycerin. Animal. 2016;**10**:1402-1408. DOI: 10.1017/S1751731116000094

[68] Edwards HD, Anderson RC, Miller RK, Taylor TM, Hardin MD, Smith SB, et al. Glycerol inhibition of ruminal lipolysis in vitro. Journal of Dairy Science. 2012;**95**(9):5176-5181. DOI: 10.3168/jds.2011-5236

[69] Lourenço M, Ramos-Morales E, Wallace RJ. The role of microbes in rumen lipolysis and biohydrogenation and their manipulation. Animal. 2010;**4**(7):1008-1023. DOI: 10.1017/S175173111000042X

[70] Maia MR, Chaudhary LC, Figueres I, Wallace RJ. Metabolism of polyunsaturated fatty acids and their toxicity to the microflora of the rumen. Antonie Van Leeuwenhoek. 2007;**91**:303-314. DOI: 10.1007/s10482-006-9118-2

[71] Schmid A, Collomb M, Sieber R, Bee G. Conjugated linoleic acid in meat and meat products: A review. Meat Science. 2006;**73**(1):29-41. DOI: 10.1016/j.meatsci.2005.10.010

[72] Wallace RJ, Chaudhary LC, McKain N, McEwan NR, Richardson AJ, Vercoe PE, et al. *Clostridium proteoclasticum*: A ruminal bacterium that forms stearic acid from linoleic acid. FEMS Microbiology Letters. 2006;**265**:195-201. DOI: 10.1111/j.1574-6968.2006.00487.x

[73] Favaro VR, Ezequiel JMB, D'Aurea AP, van Cleef EHCB, Sancanari JBD, Santos V, et al. Glycerin in cattle feed: Intake, digestibility, and ruminal and blood parameters. Semina-Ciências Agrárias. 2015;**36**:1495-1505. DOI: 10.5433/1679-0359.2015v36n3p1495

[74] Paschoaloto JR, Ezequiel JMB, Almeida MTC, Favaro VR, Homem AC, de Carvalho VB, et al. Inclusion of crude glycerin with different roughages changes ruminal parameters and in vitro gas production from beef cattle. Ciencia Rural. 2016;**46**(5):889-894

[75] Brown MS, Ponce CH, Pulikanti R. Adaptation of beef cattle to high-concentrate diets: Performance and ruminal metabolism. Journal of Animal Science. 2006;**84**:E25-E33

[76] Missio RL, Brondani IL, Alves Filho DC, Silveira MF, Freitas LS, Restle J. Comportamento ingestivo de tourinhos terminados em confinamento, alimentados com diferentes níveis de concentrado na dieta. Revista Brasileira de Zootecnia. 2010;**39**(7):1571-1578. DOI: 10.1590/S1516-35982010000700025

[77] Resende Junior JC, Alonso LS, Pereira MN, Magallanes MGR, Duboc MV, Oliveira EC, et al. Effect of the feeding pattern on rumen wall morphology of cows and sheep. Brazilian Journal of Veterinary Research and Animal Science. 2006;**43**:526-536

[78] Dirksen G. The rumen acidosis complex--recent knowledge and experiences (1). A review. Tierarztliche Praxis. 1985;**13**(4):501-512

[79] Costa SF, Pereira MN, Melo LQ, Resende JC, Chaves ML. Lactate, propionate and, butyrate induced morphological alterations on calf ruminal mucosa and epidermis—I Histologycals aspects. Arquivo Brasileiro de Medicina Veteriária e Zootecnia. 2008;**60**:1-9. DOI: 10.1590/S0102-09352008000100001

[80] Gorka P, Kowalski ZM, Pietrzak P, Kotunia A, Kiljanczyk R, Flaga J, et al. Effect of sodium butyrate supplementation in milk replacer and starter diet on rumen development in calves. Journal of Physiology and Pharmacology. 2009;**4**(5):10-11

[81] Sander EG, Warner RG, Harrison HN, Loosli JK. The stimulatory effect of sodium butyrate and sodium propionate on the development of rumen mucosa in the young calf. Journal of Dairy Science. 1959;**42**(9):1600-1605